Sociobiology
and Human Nature

*An Interdisciplinary
Critique and Defense*

Michael S. Gregory, Anita Silvers & Diane Sutch
Editors

Sociobiology
and Human Nature

 Jossey-Bass Publishers
San Francisco • Washington • London • 1978

SOCIOBIOLOGY AND HUMAN NATURE
An Interdisciplinary Critique and Defense
by Michael S. Gregory, Anita Silvers, and Diane Sutch, Editors

Copyright © 1978 by: Jossey-Bass, Inc., Publishers
433 California Street
San Francisco, California 94104

&

Jossey-Bass Limited
28 Banner Street
London EC1Y 8QE

Library of Congress Catalogue Card Number LC 78-62559

International Standard Book Number ISBN 0-87589-384-8

Manufactured in the United States of America

JACKET DESIGN BY WILLI BAUM

301.1
G823s

FIRST EDITION

Code 7824

191845

K.D

The Jossey-Bass Social and Behavioral Science Series

Preface

This volume is the outcome of a conference held at San Francisco State University on June 14–15, 1977. That conference, "Sociobiology: Implications for Human Studies," was sponsored by NEXA, the Science-Humanities Convergence Program. NEXA is funded under a development grant from the National Endowment for the Humanities (NEH). It is responsible for deriving a curriculum of courses, team taught by members of the humanities and the sciences, that stress the convergence of these "two cultures" within a context of common history and implicitly shared values. NEXA is also charged with faculty development at San Francisco State University, enabling instructors to share the perspectives of disciplines often far removed from their own. Further, under the auspices of NEH and private funding institutions NEXA is able to offer public symposia and colloquia that focus on concerns that equally affect sciences and the humanities.

The conference was designed to perform two functions for

the academic community and the general public. First, it allowed an airing of the principles and assumptions of the exciting and controversial new discipline. Second, it allowed the convergence of expert interdisciplinary opinion on the humanistic implications of sociobiological inquiry and the speculations deriving from that inquiry.

Thus, the purpose of the NEXA symposium was not to present the full theoretical background of sociobiology or to discuss exhaustively its methodological strategies. Rather, NEXA provided a setting in which biologists, sociologists, anthropologists, psychologists, physicists, economists, and humanists could combine their efforts to understand the import of the questions currently being raised in sociobiological research.

The symposium at San Francisco State University was designed and coordinated by Anita Silvers. It was divided into morning and afternoon sessions, with a concluding panel on the evening of the second day. The first panel, titled "What is Sociobiology?" was chaired by Alex C. Sherriffs, vice chancellor, academic affairs, California State University and Colleges. Panel members included: David P. Barash, Frank A. Beach, and George E. Pugh. The second panel, "Sociobiology: The Problem of the Human Mind," was chaired by Jerold M. Lowenstein, director of nuclear medicine, Presbyterian Hospital, San Francisco. Panel members included: John L. Fuller, Marjorie Grene, and Donald R. Griffin. The third panel, "Sociobiology: The Problem of Human Choice," was chaired by Virginia L. Olesen, Department of Social and Behavioral Sciences, University of California, San Francisco. Panel members included: Garrett Hardin, Karl Pribram, and John R. Searle. The fourth panel, "Sociobiology and Society," was chaired by DeVere Pentony, dean, School of Behavioral and Social Sciences, San Francisco State University. Panel members included: Kenneth Boulding, George Wald, and S. L. Washburn. The final panel, "Sociobiology: The Long View," was chaired by James C. Kelley, dean, School of Science, San Francisco State University. Panel members included: Gerald Holton, David L. Hull, and J. B. Schneewind.

Of the fifteen panelists attending the NEXA symposium,

thirteen are represented in this volume. Three additional contributions are included, and we thank Pierre L. van den Berghe, Joseph S. Alper, and Edward O. Wilson for preparing chapters. We are particularly grateful to Edward Wilson for agreeing to review all manuscripts and for providing an initial overview of the field that he has done so much to bring to academic and public attention.

We also thank all those persons, too numerous to mention here, who made both the 1977 symposium and this book possible. We must specifically acknowledge our debt to the National Endowment for the Humanities; to Paul F. Romberg, president of San Francisco State University; to Robert I. Bowman and Jack Tomlinson of our Department of Biology; to Frances Fields, administrative coordinator; and to all members of the NEXA faculty and staff for their hard work and unfailing enthusiasm.

San Francisco, California   Michael S. Gregory
August 1978   Anita Silvers
Diane Sutch

Contents

Contents

Contributors

MICHAEL S. GREGORY is professor of English at San Francisco State University, where he has taught since 1959, and director of the NEXA Program funded under a development grant from the National Endowment for the Humanities (NEH). He received his bachelor's degree (1952) in a group major (English, anthropology, and psychology) and his doctor's degree (1969) in cultural anthropology from the University of California at Berkeley.

In 1965–66, Gregory conducted field research in Hong Kong under a grant and fellowship from the National Institute of Mental Health. He was director of an NEH planning grant, "Major Figures and Their Impact," in 1973–74. This project was the precursor of the present NEXA Program, which combines the teaching of humanities and science within the framework of intellectual history. Gregory serves regularly as a panelist and reviewer of grants for six programs within the NEH and the National Science Foundation; he was a founding member of the NEH national board of consultants.

Gregory is the author of a number of articles dealing with China, strategies of teaching, and the place of humanities in a scientific world. He has also published short fiction.

ANITA SILVERS is professor of philosophy at San Francisco State University. She received her bachelor's degree (1962) from Sarah Lawrence College. After postgraduate work at London University, she obtained her doctor's degree (1967) in philosophy from Johns Hopkins University.

In June 1977, she served as coordinator for the NEXA symposium, "Sociobiology: Implications for Human Studies," and in December 1977 she coordinated the NEXA conference, "The Recombinant DNA Controversy: Public Policy at the Frontier of Knowledge."

Elected a trustee of the American Society for Aesthetics in 1976, Silvers has written about aesthetics and ethics. Her current research involves art, perception, and cognition.

DIANE SUTCH is a graduate student in philosophy at San Francisco State University and is a research technician for the NEXA Program. She served as an assistant to Marjorie Grene at the 1977 Council for Philosophical Studies Summer Institute, "Biological and Social Perspectives on Human Nature." She received her bachelor's degree in communication theory in 1976.

JOSEPH S. ALPER, associate professor, Sociobiology Study Group, Science for the People, Cambridge, Massachusetts, and Department of Chemistry, University of Massachusetts, Boston

DAVID P. BARASH, associate professor, Department of Psychology and Zoology, University of Washington

FRANK A. BEACH, professor, Department of Psychology, University of California, Berkeley

KENNETH E. BOULDING, president-elect of the American Association for the Advancement of Science and professor of economics, Institute of Behavioral Science, University of Colorado, Boulder

JOHN L. FULLER, professor, Department of Psychology, State University of New York, Binghamton

MARJORIE GRENE, professor, Department of Philosophy, University of California, Davis

DONALD R. GRIFFIN, professor, Department of Biology, Rockefeller University, New York City

GARRETT HARDIN, professor of human ecology, Department of Biological Sciences, University of California, Santa Barbara

GERALD HOLTON, Mallinckrodt Professor of Physics and professor of the history of science, Jefferson Physical Laboratory, Department of Physics, Harvard University

DAVID L. HULL, professor, Department of Philosophy, University of Illinois, Chicago Circle

J. B. SCHNEEWIND, professor, Department of Philosophy, Hunter College, City University of New York

JOHN R. SEARLE, professor, Department of Philosophy, University of California, Berkeley

PIERRE L. VAN DEN BERGHE, professor, Department of Sociology, University of Washington

GEORGE WALD, 1967 Nobel Prize in Physiology or Medicine, Higgins Chair in Biology, and professor, the Biological Laboratories, Harvard University

S. L. WASHBURN, professor, Department of Anthropology, University of California, Berkeley

EDWARD O. WILSON, Baird Professor of Science, professor of Zoology, and curator in Entomology, Museum of Comparative Zoology, Harvard University

Sociobiology and Human Nature

*An Interdisciplinary
Critique and Defense*

Edward O. Wilson

Introduction: What Is Sociobiology?

I was surprised—even astonished—by the initial reaction to *Sociobiology: The New Synthesis* (1975a). When the book was published in 1975, I expected a favorable reaction from other biologists. After all, my colleagues and I had merely been extending neo-Darwinism into the study of social behavior and animal societies, and the underlying biological principles we employed were largely conventional. The response was in fact overwhelmingly favorable. From the social scientists, I expected not much reaction. I took it for granted that the human species is subject to sociobiological analysis no less than to genetic or endocrinological analysis; the final chapter of my book simply completed the catalogue of social species by adding *Homo sapiens*. I hoped to make a contribution to the social sciences and humanities by laying out, in immediately accessible form, the most relevant methods and principles of population biology, evolutionary theory, and sociobiology. I expected that many social scientists, already convinced of the necessity of a biological foundation for their subject, would be tempted to pick up the tools and try them out. This has occurred to a limited extent, but there

1

has also been stiff resistance. I now understand that I entirely underestimated the Durkheim-Boas tradition of autonomy of the social sciences, as well as the strength and power of the antigenetic bias that has prevailed as virtual dogma since the fall of Social Darwinism.

I did not even think about the Marxists. When the attacks on sociobiology came from Science for the People, the leading radical left group within American science, I was unprepared for a largely ideological argument. It is now clear to me that I was tampering with something fundamental: mythology. Evolutionary theory applied to social systems is an extension of the great Western traditions of scientific materialism. As such, it threatens to transform into testable hypotheses the assumptions about human nature made by some Marxist philosophers. Its first line of evidence is not favorable to those assumptions, insofar as most traditional Marxists cling to a vision of human nature as a relatively unstructured phenomenon swept along by economic forces extraneous to human biology. Marxism and other secular ideologies previously rested secure as unchallenged satrapies of scientific materialism; now they were in danger of being displaced by other, less manageable biological explanations. The remarkably harsh response of Science for the People is an example of what Hans Küng (1976) has called the fury of the theologians.

But much of the confusion has come from a simple misunderstanding of the content of sociobiology. Sociobiology is defined as the systematic study of the biological basis of all forms of social behavior, including sexual and parental behavior, in all kinds of organisms, including humans. As such, it is a discipline—an inevitable discipline, since there must be a systematic study of social behavior. Sociobiology consists mostly of zoology. About 90 percent of its current material concerns animals, even though over 90 percent of the attention given to sociobiology by nonscientists, and especially journalists, is due to its possible applications to the study of human social behavior. There is nothing unusual about deriving principles and methods, and even terminology, from intensive examinations of lower organisms and applying them to the study of human beings. Most of the fundamental principles of genetics and biochemistry applied to human biology are based on colon bacteria, fruit flies, and white rats. To say that the same science can be applied to human beings is not to reduce humanity to the status of these simpler creatures.

Nor is there anything new or surprising about having such a discipline within the family of the biological sciences. The term *sociobiology* was used independently by John P. Scott in 1946 and by Charles F. Hockett in 1948, but the word was not picked up immediately by others. In 1950, Scott, who had been serving as secretary of the small but influential Committee for the Study of Animal Behavior, suggested *sociobiology* more formally as a term for the "interdisciplinary science which lies between the fields of biology (particularly ecology and physiology) and psychology and sociology" (p. 1004). From 1956 to 1964, Scott and others constituted the Section on Animal Behavior and Sociobiology of the Ecological Society of America. This Section became the present Animal Behavior Society. During 1950–1970, *sociobiology* was employed intermittently in technical articles, a usage evidently inspired by its already quasi-official status. But other expressions, such as *biosociology* and *animal sociology,* were also employed. When I wrote the final chapter of *The Insect Societies* (1971), which was entitled "The Prospect for a Unified Sociobiology" and when I wrote *Sociobiology: The New Synthesis* (1975a), where I suggested that a discrete discipline should now be built on a foundation of genetics and population biology, I selected the term *sociobiology* rather than some other, novel expression because I believed it would already be familiar to most students of animal behavior and hence more likely to be accepted.

Pure sociobiological theory, being independent of human biology, does not imply by itself that human social behavior is determined by genes. It allows for any one of three possibilities. One is that the human brain has evolved to the point that it has become an equipotential learning machine entirely determined by culture. The mind, in other words, has been freed from the genes. A second possibility is that human social behavior is under genetic constraint but that all of the genetic variability within the human species has been exhausted. Hence our behavior is to some extent influenced by genes, but we all have exactly the same potential. A third possibility, close to the second, is that the human species is prescribed to some extent but also displays some genetic differences among individuals. As a consequence, human populations retain the capacity to evolve still further in their biological capacity for social behavior.

I consider it virtually certain that the third alternative is the correct one. Because the evidence has been well reviewed in other recent works, most notably Chagnon and Irons (Eds., in press), De Vore (in press), and Freedman (in press), I will not undertake to exemplify it or review it in detail. Instead, let me outline its content.

1. *Specificity of human social behavior.* Although the variation of cultures appears enormous to the anthropocentric observer, all human behavior together comprises only a tiny subset of the realized social systems of the thousands of social species on earth. Corals and other colonial invertebrates, the social insects, fish, birds, and nonhuman mammals display among themselves an array of arrangements that it is difficult for human beings even to understand, much less imitate. Even if we were to attempt to duplicate some of these social behaviors by conscious design, it would be a charade likely to create emotional breakdown and a rapid reversal of the effort.

2. *Phylogenetic relationships.* Our social arrangements most closely resemble those of the Old World monkeys and apes, which on anatomical and biochemical grounds are our closest living relatives. This is the result expected if we share a common ancestry with these primates, which appears to be an established fact, and if human social behavior is still constrained to some extent by genetic predispositions in behavioral development.

3. *Conformity to sociobiological theory.* In the case of the hypothesis of genetic constraints on human social behavior, it should be possible to select some of the best principles of population genetics and ecology, which form the foundations of sociobiology, and to apply them in detail to the explanations of human social organization. The hypothesis should then not only account for many of the known facts in a more convincing manner than do previous attempts but should also identify the need for new kinds of information not conceptualized by the unaided social sciences. The behavior thus explained should be the most general and least rational of the human repertory, the furthest removed from the

influence of year-by-year shifts in fashion and convention. There are in fact a substantial number of anthropological studies completed or underway that meet these exacting criteria of postulational-deductive science. Among them can be cited the work of Joseph Shepher (1971) on the incest taboo and sexual roles, Mildred Dickeman (in press) on hypergamy and sex-biased infanticide, William Irons (in press) on the relation beween inclusive genetic fitness and the local set of evaluational criteria of social success in a herding society, Napoleon Chagnon (1976) on aggression and reproductive competition in the Yanomamö, William Durham (1976) on the relation between inclusive fitness and warfare in the Mundurucú and other primitive societies, Robin Fox (personal communication) on the relation of fitness to kinship rules, Melvin Konner (1972) and Daniel G. Freedman (1974, in press) on the adaptive significance of infant development, and James Weinrich (1977) on the relationship of genetic fitness and the details of sexual practice, including homosexuality.

4. *Genetic variation within the species.* By 1977, more than 1,200 loci had been located on human chromosomes through the fine analysis of biochemical and other mutations (McKusick and Ruddle, 1977). Many of these point mutations, as well as a growing list of chromosomal aberrations, affect behavior. Most simply diminish mental capacity and motor ability, but at least two, the Lesch-Nyhan syndrome, based on a single gene, and Turner's syndrome, caused by the deletion of a sex chromosome, alter behavior in narrow ways that can be related to specific neuromuscular mechanisms. The adrenogenital syndrome, which is induced by a single recessive gene, appears to masculinize girls through an early induction of adrenocortical substances that mimic the male hormone.

More complex forms of human behavior are almost certainly under the control of polygenes (genes scattered on many chromosome loci), which in turn create their effects through alternating a wide array of mediating devices, from elementary neuronal wiring to muscular coordination and "mental set" induced by hormone levels. In most instances, the role of behavioral polygenes can be evaluated—but only qualitatively—by the careful application of twin and adoption studies. The most frequently used method is to compare the similarity between identical twins, who are known to be genetically identical, with the similarity between fraternal twins, who are no closer genetically than ordinary siblings. When the similarity between identical twins proves greater, this distinction between the two kinds of twins is ascribed to heredity. Using this and related techniques, geneticists have found evidence of a substantial amount of hereditary influence on the development of a variety of traits that affect social behavior, including number ability, word fluency, memory, the timing of language acquisition, sentence construction, perceptual skill, psychomotor skill, extroversion and introversion, homosexuality, the timing of first heterosexual activity, and certain forms of neurosis and psychosis, including the manic-depressive syndrome and schizophrenia.

In most instances, there is a flaw in the results that renders most of them less than definitive: Identical twins are commonly treated more alike by their parents than are fraternal twins. They are instructed in a more nearly parallel manner, dressed more alike, and so forth. In the absence of better controls, it is possible that the greater similarity of identical twins could, after all, be due to environmental influences and not their genetic identity. However, new and more sophisticated studies have begun to take account of this additional factor. Loehlin and Nichols (1976), for example, analyzed many aspects of the environments and performances of 850 sets of twins who took the National Merit Scholarship test in 1962. The early histories of the subjects, as well as the attitudes and rearing practices of the parents, were taken into account. The results showed that the generally more similar treatment of the identical twins cannot account for their greater similarity in general abilities and personality traits or even in ideals, goals, and vocational interests. It is evident that either the similarities are based in substantial part on genetic identity or else environmental agents were at work that remained hidden to Loehlin and Nichols.

My overall conclusion from the existing information is that *Homo sapiens* is a typical animal species with reference to the quality and magnitude of the genetic diversity affecting its behavior. I also believe that it will soon be within our ability to locate and characterize specific genes that alter the more complex forms of social behavior. Obviously, the alleles discovered will not prescribe different dialects or modes of dress. They are more likely to work measurable changes through their effects on learning modes and timing, cognitive and neuromuscular ability, and the personality traits most sensitive to hormonal mediation. If social scientists and sociobiologists somehow choose to ignore this line of investigation, they will soon find human geneticists coming up on their blind side. The intense interest in medical genetics, fueled now by new methods such as the electrophoretic separation of proteins and rapid sequencing of amino acids, has resulted in an acceleration of discoveries in human heredity that is certain to have profound consequences for the study of genetics of social behavior.

I wish now to take up the concerns expressed about human sociobiology in the chapters to follow in *Sociobiology and Human Nature*. Most have been expressed by other authors in one form or another before the NEXA conference. I have no desire to rebut specific points raised by individual authors. This would in any case be unfair by the ordinary canons of debate, and *Sociobiology and Human Nature* surely is a debate. Rather, I want to discuss in broader terms the ways in which the several intellectual traditions represented so well by the other contributors might be reconciled with the relatively uncompromising biological approach I have taken up to the present time.

The first area of conflict that can be resolved is the relation of genes to culture. Many social scientists see no value in sociobiology because they are persuaded that variation among cultures has no genetic basis. Their premise is right, their conclusion wrong. We can do well to remember Rousseau's dictum that those who wish to study humans should stand close, while those who wish to study humanity should look from afar. The social scientist is interested in the often microscopic, but important, variations in behavior that almost everyone agrees are due to culture and the environment. The sociobiologist is interested in the more general features of human nature and the limitations that exist in the environmentally induced variation. He or she is especially interested in the fact that, although all cultures taken together constitute a very great amount of variation, their total content is far less than that displayed by the remaining species of social animals. By comparing the diagnostic features of human organization with those of other primate species, the sociobiologist aims to reconstruct the earliest evolutionary history of social organization and to discern its genetic residues in contemporary societies. The approach is entirely complementary to that of the social sciences and in no way diminishes their importance—quite the contrary.

Those immersed in the rich lore of the social sciences sometimes reject human sociobiology because it is reductionistic. But almost all of the great advances of science have been made by reduction, in the form of conjectures that are often bold and momentarily premature. Theoretical physics transformed chemistry, chemistry transformed cell biology and genetics, natural selection theory transformed ecology—all by stark reduction, which at first seemed inadequate to the task. Reduction is a method by which new mechanisms and relational processes are discovered. In the most successful case histories of postulational-deductive science, propositions are expressed in forms that can be elaborated into precise, testable models. The other side of reduction, the antithesis of the thesis, is synthesis. As the new principles and equations are validated by repeated testing, they are used in an attempt to reconstitute the full array of the subject's phenomena. Karl Popper (1974) has correctly suggested that philosophical reductionism is wrong but that methodological reductionism is necessary for the advancement of science. Here is how I tried to summarize the role of sociobiological reduction in an earlier review (Wilson, 1977, p. 138):

> The urge to be reductionistic is an understandable human trait. Ernst Mach [1974] captured it in the following definition: "Science may be regarded as a minimal problem consisting of the completest presentment of facts with the least possible expenditure of thought." This is a sentiment of a member of the antidiscipline, impatient to set aside complexity and get on with the search for more fundamental ideas. The laws of his subject are necessary to the discipline above, they challenge and force a mentally more efficient restructuring, but they are not sufficient for its purposes. Biology is the key to human nature, and social scientists cannot afford to ignore its emerging principles. But the social sciences are potentially far richer in content. Eventually they will absorb the relevant ideas of biology and go on to beggar them by comparison.

The strongest redoubt of counterbiology appears to be mentalism. It is difficult—for some it is impossible—to envision the existence of the mind and the creation of symbolic thought by biological processes. "The human mind," this argument often goes, "is an emergent property of the brain that is no longer tied to genetic controls. All that the genes can prescribe is the construction of the liberated brain." But the relation between genes, the brain, and the mind is only a practical difficulty, not a theoretical one. Models have already been produced in neurobiology and cognitive psychology that allow at least the possibility of mind as an epiphenomenon of complex but essentially conventional neuronal circuitry. Consciousness might well consist of large numbers of coded abstractions, some fed stepwise through a hierarchy of integrating centers whose lowest array consists of the primary sense cells, others originating internally to simulate these hierarchies. The brain—in Charles Sherrington's (1940) metaphor, the "enchanted loom where millions of flashing shuttles weave a dissolving pattern"—not only experiences scenarios fed to it by the sensory channels but also creates them by recall and fantasy. In sustaining this activity, the brain depends substantially on the triggering effect of verbal symbols. There is also a reliance on what have been called *plans* or *schemata*—configurations within the brain, either innate or experiential in origin, against which the input of the nerve cells is compared. The matching of the real or expected patterns can have one or more of several effects. It can contribute to mental "set," the favoring of certain kinds of sensory information over others. It can generate the remarkable phenomena of *gestalt* perception, in which the mind supplies missing details from the actual sensory information in order to complete a pattern and make a classification. And it can serve as the physical basis of will: The mind can be guided in its actions by feedback loops that lead from the sense organs to the brain schemata to the neuromuscular machinery and sense organs and back again until the schemata "satisfy" themselves that the correct action has been taken. The mind could be a republic of alternative schemata, programmed to compete for control of the decision centers, individually waxing and waning in power according to the relative urgency of the needs of the body being signaled through other nervous pathways passing upward

through the lower brain centers. The mind might or might not
work approximately in such a manner. My point is that it is entirely
possible for all known components of the mind, including will, to
have a neurophysiological basis subject to genetic evolution by natural selection. There is no *a priori* reason why any portion of the
foundation of human social behavior must be excluded from the
domain of sociobiological analysis.

Some critics have objected to the drawing of analogies between animal and human behavior, especially as it entails the same
terminology to describe phenomena across species. This reserva-
tion has always struck me as insubstantial. The definitions and limitations of the concepts of analogy and homology have been well
worked out by evolutionary biologists, and it is difficult to imagine
why the same reasoning cannot be extended with proper care to
the human species. We already speak of the octopus eye and the
human eye, insect copulation and human copulation, and earth-
worm learning and human learning, even though in each of these
cases the two species are in different superphyla, and the traits
listed were independently evolved. The questions of interest are
in fact the degrees of convergence and the processes of natural
selection that made the convergence so close. When biologists compare altruism in the honeybee worker with human altruism, no one
seriously believes that they are based on homologous genes or that
they are identical in detail. Slavery practiced by *Polyergus* and *Stron-
gylognathus* ants resembles human slavery in some broad features
and differs from it in others, as well as in most details of its execution. By using the same term for such comparisons, the biologist
calls attention to the fact that some degree of convergence has occurred and invites an analysis of all the causes of similarity and
difference. There is a Greek-derived term for insect slavery—*du-
losis*—but its usage outside entomology would not only complicate
language but would also slow the very comparative analysis that is
of greatest interest.

I am most puzzled by the occasional demurral that sociobiology distracts our attention from the real needs of the world.
The questions are raised, "How can we worry about the origins of
human nature when the nuclear sword hangs over us? When people are starving in the Sahel and in Bangladesh and political pris-

oners are rotting in Argentinian jails?" In response, one can answer, "Do we want to know, in depth and with any degree of confidence, why we care? And, after these problems have been solved, what then?" The highest goals professed by governments everywhere are human fulfillment above the animal level and the realization of individual potential. But what is fulfillment, and to what ends can potential be expanded? I suggest that only a deeper understanding of human nature, which must be developed from neurobiological investigations of the brain and the phylogenetic reconstruction of the species-specific properties of human behavior, can provide humanity with the perspective it requires to formulate its highest social goals. '

The excitement of sociobiology comes from the promise of the role it will play in this new humanistic investigation. Its potential importance beyond zoology lies in its logical position as the bridging discipline between the natural sciences on the one side and social sciences and humanities on the other. For years, the chief spokespersons of the natural sciences to Western high culture have been physicists, astronomers, geneticists, and molecular biologists—articulate and persuasive scholars whose understanding of the evolution of the brain and of social behavior was unfortunately minimal. Their perception of values and the human condition was almost entirely intuitive and hence scarcely better than that of other intelligent laypersons. Biology has been employed as a science that accounts for the human body; it concerns itself with technological manifestations such as the conquest of disease, the green revolution, energy flow in ecosystems and the cost-benefit analysis of gene splicing. Natural scientists have by and large conceded social behavior to be biologically unstructured and hence the undisputed domain of the social sciences. For their part, most social scientists have granted that human nature has a biological foundation, but they have regarded it as of marginal interest to the resplendent variations in culture that hold their professional attention.

In order for the fabled gap between the two cultures to be truly bridged, social theory must incorporate the natural sciences into its foundations, and for that to occur biology must deal systematically with social behavior. This competence is now being ap-

proached through the two-pronged advance of neurobiology, which
boldly hopes to explain the physical basis of mind, and sociobiology, which aims to reconstruct the evolutionary history of human
nature. Sociobiology in particular is still a rudimentary science. Its
relevance to human social systems is still largely unexplored. But
in the gathering assembly of disciplines it holds the greatest promise of speaking the common language.

1

David P. Barash

Evolution as a Paradigm for Behavior

$\mathcal{A}$ revolution is underway in the study of behavior. This is certainly true of *animal* behavior and may have implications for *human* behavior as well. In fact, this symposium is testimony to the growing interest and concern over such implications. Like most revolutions, this one began when existing conditions were seen as inadequate and a presumed better way had been identified. Like most revolutions, this one has generated great excitement among the participants and equivalent resistance from the "establishment." And, like most revolutions in progress, there is no telling whether it will be successful and, if so, where it ultimately will lead. This chapter is intended partly as a manifesto for that revolution but even more as a review of its historical antecedents, its methods of operation, and its scientific assumptions, so that observers as well as participants can appreciate the basic issues. Finally, I will attempt

This manuscript was prepared while I was a fellow at the Center for Advanced Study in the Behavioral Sciences, Stanford, California. I would like to thank the staff, my colleagues, and most especially the director, Gardner Lindzey, for generating an ideal intellectual and personal environment.

to clarify and respond to some of the controversy and to prophesy future directions. We may or may not eventually agree about its moral, ethical, political, and/or philosophical implications, but I hope we can all *begin* by agreeing as to what the revolution *is* and what it is *not*.

This revolution derives essentially from the application of evolutionary biology to social behavior—the name *sociobiology* has been applied to this complex of data and theory (Wilson, 1975a; Barash, 1977). Previously, the study of animal behavior, both human and nonhuman, had lacked a consistent underlying framework—a "paradigm," in the sense of Kuhn (1962). Of course, there have been many efforts at recognizing a single, coherent paradigm for behavior, many of them reminiscent of Mark Twain's famous observation that it was easy to stop smoking—he had done it many times! Students of animal behavior have rushed frenetically to sit at the feet of one guru after another, ever hoping to discover generalized enlightenment, and ever disappointed. For example, great enthusiasm was generated sequentially, first by Loeb's theory of forced movements, then by Sherrington's unraveling of the reflex arc, Pavlov's discovery of conditioned reflexes, and the refinement of behavior modification by Thorndike and, more recently, Skinner. But, although these efforts have been variously successful in providing techniques for the *manipulation* of behavior, they have generally failed to *explain* very much, especially the complex behaviors of free-living animals.

Given that the behavior of nonhuman animals has effectively resisted generalization, it should not be surprising that the behavior of *human* animals, with its greater complexity, has also remained beyond our firm grasp. Psychoanalytical and psychological theories have proliferated in nearly direct proportion to the abundance of their practitioners; we have Freudians, Jungians, Adlerians, Skinnerians, and Piagetians, to name just a few. The social sciences concerned with group phenomena have generated equivalent Towers of Babel, and both sociology and anthropology are filled with the disciples of Durkheim, Lévi-Strauss, Spencer, Marx, and numerous other social theorists, whose ideas and approaches are often independent, mutually exclusive, and disquietingly noncomplementary, as well as resistant to verification or refutation. On the one hand, this diversity of approach may be entirely appropriate to the unique complexity of the subject, and perhaps it should be encouraged rather than bemoaned. On the other hand, internal coherence is one of the distinguishing characteristics of good science, and its absence may well have contributed to the inferiority complex often suffered by the behavioral sciences relative to the physical sciences and, indeed, to biology as well. This is not to suggest that sociobiology should ultimately replace the approaches mentioned; however, sociobiology just may exhibit a strength not displayed by any of its predecessors considered alone.

The great texts of physical science, such as Pauling's *General Chemistry* (1970) and Feynman's *Lectures on Physics* (Vol. 1, 1963; Vol. 2, 1964; Vol. 3, 1965), are particularly impressive for the coherent intellectual momentum with which they develop theory and interpret empirical findings. In recent years, biology has rediscovered its own coherent paradigm: evolution by natural selection. This is reflected in the recent appearance of excellent, integrated approaches to the life sciences, notably Keeton's *Biological Science* (1972), and Simpson and Beck's *Life* (1965). Just as evolution by natural selection unifies structure, function, and past history, it may now bring behavior within the fold as well. This is the underlying premise and hope of today's sociobiology—a unifying paradigm for animal social behavior.

Before sociobiology, efforts by *biologists* to deal with behavior mainly centered on *ethology*, a discipline whose orientation was distinctly more evolutionary than that of the social sciences but less than sociobiology is today. Thus the imprint of evolutionary thinking is apparent in ethology's emphasis on careful description and observation rather than on manipulation, as well as in the importance attributed to studying free-living animals in natural or seminatural environments. Ethologists seek to determine "what evolution hath wrought," but their science is only *quasi*-evolutionary in that natural selection is rarely *used* as an analytic tool. By contrast (as we shall see later), sociobiology is totally immersed in evolution; it constantly comes back to the evolutionary process for hypotheses, predictions, and interpretations.

Historical Antecedents

Since the sociobiological revolution is currently in progress, our lack of perspective makes it difficult to identify precisely the factors that have produced it. Perhaps this will be a task for some future historian or sociologist of science, once the dust has settled. Nonetheless, some formative influences can be recognized:

Group Selection and Individual Selection. In 1962, ecologist V. C. Wynne-Edwards produced a massive book, *Animal Dispersion in Relation to Social Behavior,* presenting biologists with a startling new thesis: that animal social behavior serves as a homeostatic (negative feedback) mechanism whose function is the regulation of population size below the carrying capacity for each species in its particular environment. Although intellectually appealing, this system presupposes a high level of reproductive restraint on the part of those individuals who *altruistically* curtail their own reproduction for the benefit of the *group.* Wynne-Edwards explained this apparent difficulty by postulating "group selection," whereby altruistic behaviors were selected because their presence conferred sufficient benefit on the *group* (of which the altruist was a member) to compensate for the reduction in Darwinian fitness suffered by each *individual* altruist.

These ideas were controversial. Biologists were especially bothered by the probability that individual selection acting against particular traits *within* each group would be likely to overwhelm the effects of selection operating *between* groups. Indeed, classical Darwinian theory had been based on the differential reproduction of *individuals,* notwithstanding occasional sloppy lapses into considerations of "benefit to the species." Wynne-Edwards' challenging ideas forced an extensive reassessment of the role of individual selection in evolution (Wiens, 1966; Williams, 1966, 1971), prompted a scientific soul-searching that clarified natural selection as a process, focused attention on the adaptations of individuals, and renewed confidence in the power of selection operating on differentials in the reproductive success of individuals. More recently, several computer models have been developed, demonstrating that group selection is theoretically feasible, but only under conditions so restrictive that they are unlikely to occur in nature (Levins, 1970; Boorman and Levitt, 1972, 1973; Gadgil, 1975).

Kin Selection. Shortly after Wynne-Edwards' book appeared, geneticist W. D. Hamilton (1964) published a milestone paper on the evolution of social behavior, especially in social insects. He considered a long-standing paradox in evolutionary biology, first noted by Darwin: Worker bees, wasps, and ants are sterile, laboring for the reproductive success of the queen. Such extreme altruism appears to run counter to the expectations of natural selection, in that individuals should be strongly predisposed to reproduce *themselves.* Hamilton showed that worker sterility is explicable through the peculiar haplodiploid genetic system of the Hymenopteran insects, as a result of which workers share more genes with a sister (whom they help to rear) than they would with their own offspring, if they were to reproduce. The emphasis here is on genes rather than on individuals *per se.* Even care of offspring is only a special case of behaviors selected because they lead to maximum representation of genes coding for those behaviors. The theory of *kin selection* was thereby developed, in which inclusive fitness replaced Darwinian fitness as the evolutionary currency with which natural selection evaluates any phenotype. Inclusive fitness is the sum of individual fitness and genetic representation through relatives, with the latter devalued in proportion as relatives are more distantly related—that is, as fewer genes are shared. (Actually, story has it that the great British biologist and all-purpose genius J. B. S. Haldane prefigured the whole issue one day almost thirty years ago. While at a pub, he was asked whether he, as an evolutionist, would give up his life for his brother. His reply: "For one brother? No. But for three brothers, yes . . . or nine cousins!")

Although alternative explanations exist for insect eusociality (Lin and Michener, 1972; Alexander, 1974), Hamilton's kinship theory has received substantial empirical verification (Trivers and Hare, 1976; but see also Alexander and Sherman, 1977). It has also emerged as a major bulwark of sociobiological theory, applied to vertebrates as well as invertebrates (West-Eberhard, 1975). Kin selection has been successfully applied to behaviors as diverse as mating systems (Maynard Smith and Ridpath, 1972; Bertram,

1976), rearing systems (Brown, 1974), and alarm calling (Maynard Smith, 1965; Sherman, 1977) and it may well provide a coherent biological theory for nepotism in animals, human as well as non-human (van den Berghe and Barash, 1977).

   The Adaptive Nature of Social Systems. Concern with the linkage between behavior and evolution had long centered around simple, discrete, and rather stereotyped behaviors ("fixed action patterns" to the ethologists). With the increased popularity and success of long-term field studies of free-living animals came the growing realization that each social system is remarkably adapted to the environment(s) experienced by each species. Distinct patterns began to emerge, with regularities occurring as a function of each species' unique ecology and phylogenetic history. A major catalyst was the identification of individual female choice as a primary criterion in determining animal mating systems (Orians, 1969). Actually, this insight was itself largely due to the increased emphasis on maximization of individual fitness following the work of Hamilton (1964) and G. C. Williams (1966). Finally came the growing epiphanic insight that such regularities suggest the operation of natural selection on complex patterns of social behavior (Crook, 1970). As field studies of animal behavior became increasingly abundant and long term, they continued to support the legitimacy of this approach. Animal social systems, once a kaleidoscopic jumble of behavioral diversity, were finally seen as merely variations on the general theme of evolution by natural selection operating on individuals and generating social patterns that maximize the fitness of each.

   Natural Selection and Social Theory. Along with appreciation of the cogency of natural selection operating on individuals, the power of kin selection as an explanatory tool, and empirical confirmation via adaptive regularities in animal social systems, have come numerous extrapolations of evolutionary biology to social behavior. This diverse theoretical flowering has been especially notable in the areas of reciprocal altruism (Trivers, 1971); parental investment (Trivers, 1972); parental ability to vary the sex of offspring (Trivers and Willard, 1973); parent-offspring conflict (Trivers, 1974); spatial arrangements of individuals within a group, both territorial (Brown, 1964) and social (Hamilton, 1971); strategies of agonistic behavior (Maynard Smith, and Price, 1973); the possibility that parents manipulate the behavior of their offspring for parental advantage (Alexander, 1974); and behavioral predispositions as a function of variations in genetic relatedness (Barash, Holmes, and Greene, in press), to mention just a few. Research of this sort cries out for empirical evaluation, but this is probably less a weakness than it is a sign of vigor, indicating a youthful, aggressively expanding science.

The New Synthesis. Finally, it remained for zoologist E. O. Wilson (1975a) to bring together much of the data and theory in an encyclopedic compendium of sociobiology that continues to attract attention and generate controversy. This volume has functioned very much as did Julian Huxley's *Evolution: The Modern Synthesis* (1942), in that it brought together a great deal that had been "in the air" scientifically and that was in need of consolidation.

Revolutions are never just the work of revolutionaries alone; their way is paved by numerous circumstances making the upheaval possible, if not inevitable. Similarly, sociobiology has not developed from the above-mentioned sociobiologists alone. The intellectual climate, as generated by advances in collateral fields, also contributed to the sociobiological revolution. In particular, behavior genetics emerged rapidly as a legitimate discipline (Ehrman and Parsons, 1976), thereby providing confidence as to the legitimacy of relating evolutionary processes to behavior. In addition, evolutionary ecology has matured as a distinct approach to natural systems (Emlen, 1972; Pianka, 1974), providing increased sophistication in the use of cost-benefit analyses and optimization theory (Cody, 1974).

Methods of Operation

Biologists and especially psychologists are likely to be uncomfortable with the way sociobiology functions. In addition to their relative unfamiliarity with the use of natural selection as a guiding principle, most students in the life sciences answer questions about behavior in terms of *proximal* rather than *distal* causation. Thus, in attempting to answer the basic question of why an animal shows a particular behavior, behavioral scientists typically concern themselves with questions regarding eliciting stimuli, the role of prior experience or other developmental processes, physiological mechanisms, and so on. Such proximal issues are related to the immediate performance of the behavior in question. By contrast, an evolutionary approach to behavior is concerned with distal causation and, in particular, with the question "What is the adaptive significance of the behavior?" Sociobiologists seek the evolutionary bases for social behavior; they attempt to unravel the factors that have selected for the particular proximal mechanism that operates in each case. Another way of viewing this dichotomy is in terms of tactics versus strategy: Various tactical decisions are typically made in pursuit of some long-term strategic goal. In the case of sociobiology, the strategic goal is considered to be maximization of fitness; tactics employed toward that end may include a wide array of hormonal, neural, and developmental processes, including (depending on the species and circumstances) susceptibility to early experience, individual learning, and social traditions.

We can identify four basic ways of employing evolution in the study of social behavior: historical, evaluative, correlational, and predictive. They are presented here in chronological order of discovery, which also corresponds to a ranking of increased scientific power.

Historical. The oldest and least powerful sociobiological use of evolution, the historical, involves the attempted reconstruction of behavioral phylogenies. Of course, since behavior does not fossilize, research of this sort is limited to careful description of extant species, leading to informed speculation as to evolutionary antecedents. An unfortunate weakness of this approach is the impossibility of verifying or refuting hypotheses. Nonetheless, excellent and convincing results have been obtained, notably with studies of empid flies (Kessel, 1955), sand wasps (Evans, 1966), and fiddler crabs (Crane, 1943) among invertebrates; and gulls (Tinbergen, 1959) and ducks (Lorenz, 1958) among vertebrates. Typically, studies of this sort proceed by describing the behavior of numerous closely related species; historical insight is then gained by assuming that the ancestral condition is most closely reflected in behavior shared by the largest number of extant species, as well as in the simplest and the least specialized behaviors. These three criteria coincide in most cases, generating some confidence in the results within the limitations of the technique itself.

Evaluative. Instead of asking "What is the history of behavior in a population?" (the historical approach), we can ask "What is its adaptive significance?" This approach seeks to evaluate a behavior by how it contributes to the fitness of animals performing it; that is, for any phenotype (behavior included) to be positively selected, it must carry with it more advantages than disadvantages—or else alternative behaviors, coded by alternative alleles, would be selected. Classic examples of the evaluative approach in sociobiological research are the work of Tinbergen and his students on gull behavior (Kruuk, 1964; Patterson, 1965; Tinbergen, 1967).

Despite the obvious cogency of such research, there are problems as well, notably the danger of confusing a *personal* evaluation for the evaluation purportedly conducted by natural selection itself. It is easy to mistake one's own intellectual handiwork for that of evolution; for example, witness the turn-of-the-century observation that flamingos were pink because of the adaptive advantage conferred by blending into the sunset! Nonetheless, evaluating the adaptive significance of behavior is an important and rewarding endeavor, and external confirmation is often available. In some cases, competing hypotheses as to the precise adaptive significance of behavior can be evaluated directly in the field (Hoogland and Sherman, 1976), and comparisons of different but closely related species may also help. For example, Tinbergen's (1963) evaluation of eggshell removal by gulls as an antipredator strategy was greatly strengthened by the discovery that cliff-nesting kittiwakes, lacking predators, also lacked this behavior (Cullen, 1957).

Correlational. Adaptive significance of behavior is evaluated by considering the ecology of each individual and assessing the extent to which the behavior in question contributes to fitness, given the environment in each case. If behavior is specifically adapted to particular environments, then different environments should select for different behaviors. Confidence in our initial evaluation of adaptive significance is therefore greatly enhanced when consistent correlations can be identified between behavior and the relevant environmental parameters. Progress along these lines has been made in studies correlating environment and social behavior in blackbirds (Orians, 1961), weaverbirds (Crook, 1964), marmots (Barash, 1974a), African bovids (Jarman, 1974), wild sheep (Geist, 1974) and, to some extent, nonhuman primates (Eisenberg, Muckenhirn, and Rudran, 1972).

Again, as with the historical and evaluative approaches, verifying and/or falsifying the results of correlational research can be problematic. Any consistent pattern of behavior-environment matching can be interpreted as adaptive and therefore as confirming the approach. This potential circularity can be ameliorated by prediction; that is, once a correlational series is established, it is sometimes possible to predict the pattern of social behavior to be found in a given environment. Research of this sort has been conducted on marmots, in which the predicted behavior-environment correlations were found for high-elevation hoary marmots and intermediate-elevation yellow-bellied marmots (Barash, 1973b, 1974b). Similarly, Galapagos swallow-tailed gulls, whose environment is in many ways intermediate between ground-nesting gulls and the cliff-nesting kittiwakes, show behavioral adaptations that are predictably intermediate as well (Hailman, 1964).

Predictive. Just as confidence in the evaluative approach is enhanced by correlational support and the approach is bolstered by predictive success, a predictive approach is the newest and most powerful technique in sociobiology's arsenal. Its application relies on what may be called the *central theorem of sociobiology* (Barash, 1977): Insofar as the behavior in question is genetically influenced, animals ought to behave so as to maximize their inclusive fitness. Predictive tests of the central theorem have been successfully employed in studies of the chronology of nest defense in alpine accentors (Barash, 1975b), variations in paternal behavior in hoary marmots (Barash, 1975a), and the male response to apparent female adultery among mountain bluebirds (Barash, 1976a). The possibility of greater analytic precision exists through the combination of optimality modeling with sociobiological prediction, as has already been accomplished for the foraging behavior of finch flocks in the Mohave Desert (Cody, 1971) and of African weaverbirds (Katz, 1974).

Scientific Assumptions

There are two, hitherto unspoken, assumptions that underlie the practice of sociobiology. One has already been mentioned: the central theorem of sociobiology. The other is perhaps even more fundamental: All phenotypes, at every stage of ontogeny, result from the interaction of genotype and environment. This interaction principle applies to all phenotypes, with behavior a special case only because of its unique complexity and occasional lability. Natural selection can operate only on those phenotypes that represent some component of genotype; that is, differential reproduction of certain individuals will have *no effect* on the distribution reproduction of a particular trait in a population if individuals do not differ genetically with regard to their predispositions for generating that trait. Phenotypic variability due entirely to the idiosyncratic experiences of each individual provides no substrate for evolutionary change. This was first demonstrated by Johannsen's now classic inability to select for increased size among highly inbred beans and has been confirmed by numerous studies ever since. In fact, Fisher's (1930) formulation implies that the rate of evolutionary change is proportional to the intensity of selection and the genotypic variance present in the selected population.

Biologists have long appreciated that structure derives from the interaction of genotype and environment. However, difficulties often arise in conceiving a genetic substrate for behavior, despite the fact that in principle the two are identical: In both cases, DNA provides a blueprint for the elaboration of enzymes and other proteins, which ultimately are responsible for establishing the basic patterns of morphology and physiology of each organism. There is indeed a wide conceptual gap between the physical and the psychic, but it is nonetheless axiomatic among neurobiologists and physiological psychologists that behavior arises from the structure and function of the nervous system. DNA specifies neuronal organization in exactly the same way as it does the organization of bone, muscle, or gland cells. Accordingly, the role of DNA in behavior is no more incongruous than the role of DNA in structure. The only difference is in the degree of flexibility: Behavior tends to be more labile than structure, but the difference is quantitative rather than qualitative.

With the recognition that behavior follows the interaction principle comes a long-overdue synthesis of ethology and comparative psychology (Hinde, 1970; Lehrman, 1970). The old nature-nurture controversy has effectively been buried; we now realize that neither genotype nor environment, acting alone, produces behavior. Consider behavior as analogous to the sound produced by a drummer playing drums: The sound is not the product of the drums (genotype) alone; but neither is it produced by the drummer (environment) alone. Rather, it arises from the interaction of the two. Change the drums or the drummer, and the sound changes as well. Continuing the analogy, the *differences* between the sounds produced by different drums (genotypes) exposed to identical drummers (environments) are due to the differences between the drums, just as differences between the behavior of different species in identical environments are attributable to the genotypic differences between the species but not to the genotype of either, acting alone. A similar argument applies to switching drummers on the same drum (different environments experienced by the same genotype can induce different behaviors). In either case, the message should be clear: No behavior is produced by genotype acting alone, and, similarly, no behavior is produced by environment alone. The interaction principle legitimizes sociobiology in that it argues for a finite role of genes in mediating behavior. (Of course, it also argues equally strongly for a finite role of environment, and this is not inconsistent. Sociobiology does not require genetic determinism of behavior, only a genetic influence larger than zero.)

Since evolution by natural selection is generally acknowledged to be the primary factor determining the distribution of genes in a population, evolution is accordingly an appropriate paradigm for behavior—even for social behavior. Finally, sociobiology employs the central theorem both to generate additional theory and to suggest hypotheses that are amenable to empirical testing.

Areas of Controversy and Difficulty

Like all revolutions, sociobiology has met considerable resistance (Allen and others, 1976). The objections can be divided for convenience into two groups, political and scientific. Some of the political objections and a brief response to each are summarized as follows (interestingly, nearly all the controversy centers around extrapolations of sociobiology to human behavior).

1. "Sociobiology is racist and sexist and represents a return to Social Darwinism." Insofar as sociobiology assumes genetic influences on behavior, this concern is understandable. However, there have been no sociobiological treatments of racial issues. Furthermore, with its emphasis on underlying biological "universals," ostensibly shared by all *Homo sapiens* from Bushman to Wall Street businessman, sociobiology appears instead to be a cogent antidote to racism. Yet much sociobiological theory relies heavily on male-female differences in behavior; and, indeed, if the identification of evolved behavioral differences between males and females is sexist, then so is sociobiology—but so is "Mother" Nature! Alternatively, the term *sexism* seems more appropriately reserved for preferential societal treatment or valuation of one sex over the other, and sociobiology is innocent on this score. Like the *yin* and *yang* of Taoism, an evolutionary approach to sexuality recognizes female and male as complementary forms, each defining the other and achieving unity. Neither is better; both are essential, and both "win" every time a new generation is successfully produced. The specific implications for male-female differences in human behavior are unclear and, indeed, hardly explored.

As for Social Darwinism, it was an unfortunate misstatement of evolutionary biology, due especially to efforts by successful turn-of-the-century capitalists to justify their exploitation of others, including imperialism and colonialism. It was an effort in which social science readily joined and for which it has been doing penance ever since by taking a virtually monolithic stand behind environmental determinism of all behavior. Social Darwinism was a travesty of biological theory, based on the mistaken notion that "survival of the fittest" referred solely to the outcome of aggressive competition. We now recognize that natural selection proceeds by differential reproduction, and Social Darwinism accordingly has no place in sociobiology or, indeed, anywhere else.

2. "Sociobiology is a doctrine of biological (genetic) determinism." This formulation confuses genetic determinism with *genetic influences* on behavior (Dobzhansky, 1976). Behavior can be genetically influenced while still retaining a wide range of flexibility. Indeed, it is no less "genetic" if susceptible to environmental influences—notably (in humans) early experience and social learning. Geneticists know that phenotypes are not magically enclosed within a gene, like the tiny homunculi of medieval medicine. Rather, genes code for a range of potential phenotypes; they are blueprints of varying flexibility, depending on the species and the behavior in question. The difference between genetic determinism and genetic influence is like the difference between shooting a bullet at a target and throwing a paper airplane—on a windy day. Julian Huxley has warned biologists against "nothing but-ism," the mistaken notion that since humans are animals we are nothing but animals. His warning is well taken, but it can also be turned around with equal cogency: Even though humans are unique in their behavioral flexibility, it is inaccurate to assume that they are "nothing but" *tabulae rasae,* blank slates on which experience can write as it will. Nonetheless, genetic determinism is a red herring, nowhere implied in the interaction principle or elsewhere in sociobiology.

Accordingly, sociobiology does not necessarily imply that humans have surrendered their autonomy to the despotism of DNA. And yet, insofar as genes influence our behavior, inclining us to do what maximizes our fitness, sociobiology may provide real insights to the deep structure that underlies our everyday behavior. It may tell us nothing of why we chose a red tie today instead of a blue one, but it may say a great deal about why we choose to adorn our bodies in the first place. Similarly, our choice of particular presidential candidates may say much about our inclinations to have leaders at all. Does evolution mean the death of free will? All scientific theories of behavior must of necessity be relatable to natural laws, and certainly conditioning theory does not offer our species any greater freedom or dignity (Skinner, 1971). I suggest that free will is at a maximum when individuals are able to behave in accordance with their inclinations, whether these derive from early experience, social learning, cultural traditions—or evolutionary biology. Sociobiology is simply a concerned effort to understand the contribution of natural selection to behavior.

3. "Sociobiology implies support for the status quo." The idea is that an evolutionary perspective implies a Panglossian view that "all is for the best in this best of all possible worlds." There are several misunderstandings here. First, this view suggests that what is "biological" is necessarily "good" as well. The goal of sociobiology is to better understand behavior, not to legitimize our foibles. Similarly, we study pneumococci to understand them better, not because we approve of pneumonia! (A similar argument can be applied to sociobiology's purported sexism.) Second, the conception of behavior as adaptive does not imply perfection in any system; in fact, it implies quite the opposite: Evolution is notoriously opportunistic, often selecting short-term success at the cost of long-term fitness. The eventual consequence, of course, is extinction. Sewall Wright's (1969) image of "adaptive landscapes" further emphasizes the less-than-perfect nature of adaptions: Populations may get stuck on minor adaptive peaks, with the constituent individuals unable to achieve greater fitness because they cannot cross a valley separating them from a higher peak. Wright's conclusion was that a certain frequency of random movements (genetic drift), combined with the blind upward momentum of selection, is necessary for adaptation. The moral for sociobiology and its detractors is that nothing in nature is perfect. The existence of a social system is *de facto* evidence that it is at least minimally adaptive. However, this does not imply that something better cannot be found.

Finally, evolution is by its nature a doctrine of change. For that reason, in fact, it was strongly resisted by the early nineteenth-century European establishment, which had recently been shaken by the American and French revolutions and was understandably apprehensive of ideas that the natural world was mutable. Such notions tended to undermine the establishment's own claim to special position by virtue of divine will and the rigid structuring of nature implied in the doctrine of special creation. It would therefore be ironic if an evolutionary paradigm for behavior should now be seen as supportive of establishment values!

4. "Culture is so important to human behavior that genetic factors (and hence sociobiology) may be considered irrelevant." It is undoubtedly true that genetic factors are less influential in the behavior of *Homo sapiens* than they are for any other species. But this does not mean they are irrelevant. Anthropologists and sociologists often point to the enormous cultural diversity shown by our species. But that is just the point. Insofar as certain cross-cultural universals emerge as truly *human* traits, despite the enormous cultural overlay, then the evidence for biological underpinnings is all the more cogent. Cultural diversity is like the icing on the cake. Thus far, anthropologists in particular have studied the icing, while psychologists and psychiatrists have concerned themselves with how the icing is put together. Sociobiology urges us to look at the human *cake* that underlies the icing—at the basic human patterns of behavior toward kin versus nonkin, parent-offspring conflict, male-female differences in reproductive strategies, and the fundamental organization of human families. Because of ethical constraints, our hands are tied concerning direct experimentation on the genetic basis of human social behavior. But we can use sociobiology to predict the existence of certain cross-cultural universals. And then, if the shoe fits. . . .

The following scientific criticisms of sociobiology may well be more cogent than the preceding political concerns. They warrant further attention from both opponents and proponents of the new revolution.

1. "There is no evidence for specific genes influencing specific behaviors in humans beyond simple inherited metabolic and structural traits." This is true. There is abundant evidence for nonhuman animals, but, since no experimental manipulation can be performed on humans, we are reduced to extrapolation and speculation. But, unless we are willing to posit a qualitative discontinuity between the biology of human and of nonhuman animals, the interaction principle should apply to *Homo sapiens* as well. Nonetheless, it will be very difficult to disentangle biological from cultural factors in the etiology of any given human behavior, since in many cases both processes lead to the same phenotype. For example, among certain Eskimo societies old people are the first to commit suicide during famine. This is consistent with sociobiolog-

ical theory, since inclinations toward altruism should increase as personal reproductive potential decreases. It is also explicable by kin selection, group selection, parental manipulation, and even reciprocal altruism, extended across generations. But geriatric self-sacrifice could also be due to cultural traditions that celebrate people who behaved that way in the past. In a sense, successful cultural traditions are those that have withstood a process analogous to natural selection (Campbell, 1975); accordingly, they tend to be adaptive, at least for societies, if not for individuals. But, unlike organic evolution, cultural evolution is Lamarckian, not Darwinian. However, the results of these two very different phenomena can be disquietingly similar and will be a challenge to assess.

2. "Natural selection is not the only factor responsible for the distribution of genes in a population; in particular, stochastic processes may be important." This also is true. At present, however, most evidence supports the role of selection as the primary, although not sole, determinant of gene distribution (Ayala, 1975). Furthermore, stochastic processes can be modeled and included in sociobiological theory—in certain areas, a start has already been made (Cohen, 1971).

3. "Sociobiology explains everything. It is too easy to generate untestable theory." Again, this is a serious objection. It is easier to produce theory than to test it—in fact, this is a continuing problem with other aspects of evolutionary theory as well (Peters, 1976). But this difficulty is not insurmountable; in particular, adroit use of the predictive approach should keep sociobiology on a firm empirical footing. There are additional, related problems. For example, we simply do not know how much "adaptiveness" genes can store. Theory suggests that individuals should behave toward each other as a function of cost-benefit fitness considerations, in which genetic relatedness figures prominently. Indeed, much theory turns on the ability of animals to respond to subtle differences in fitness considerations; yet we do not know whether most behavioral predispositions are literally capable of such fine tuning: "How do I love thee? Let me count thy genes." Still, by making the expectations of theory as explicit as possible, they can eventually be susceptible to confirmation or refutation, just as with any science (Barash, 1976b).

In the current flush of theoretical enthusiasm, there is certainly a danger that sociobiologists are constructing unsupported bridges, until, like cartoon characters, they may look down one day only to discover they are building on air! With a discipline so ripe for theory and simultaneously so resistant to the acquisition of pertinent data, such expeditions are perhaps inevitable. The sociobiological imagination may ultimately require weights rather than wings, but even now there are enough hard data to restrain excessive light-footedness.

Future Directions

As to the future, some limited predictions can be made with confidence. There will certainly be further theoretical insights and, hopefully, vigorous attempts to consolidate past gains by gathering appropriate data on free-living populations. Sociobiology may be about to enter a period of "normal science" (Kuhn, 1962) in which the dizzying advances of the past ten years will be evaluated, solidified, and perhaps extended into new realms. We hope to learn the precise adaptive limits of which organisms are capable; that is, it is one thing for theorists to suggest behavioral tactics that would ultimately serve the strategic goal of fitness maximization—it is another to see whether animals *really do* behave optimally. Studies assessing the predicted correlation between altruism and genetic relatedness, for example, would go far toward this goal, and accurate field techniques permitting measurement of genetic relatedness between individuals would be extremely valuable. On the one hand, sociobiology will almost certainly expand with further extrapolations to human behavior, perhaps providing a new perspective on the data of anthropology and the theories of sociology. Kin selection in particular may provide a means of evaluating human family structure in biological terms (van den Berghe and Barash, 1977). I doubt very much that evolutionary biology, in itself, will provide ethical or moral guidelines. On the other hand, it is very likely that new light will be thrown on the bases of our own human morality. The Golden Rule, incest taboos, patterns of inheritance and of interaction between old and young—all these are interpretable through natural selection. And the end is not in sight. Indeed, we are barely glimpsing the beginning!

Prophesying more boldly now, the evolutionary paradigm may provide considerable fresh insight to old problems. For example, sociobiology's concern with genetic relatedness suggests a new perspective on multicellular organisms, such as that their physical and functional integrity may be due to the genetic identity of the participating cells. (Why else would liver cells labor altruistically to detoxify dangerous substances, leaving reproduction to the gonadal cells?) In this sense, senescence may be explicable as the necessary consequence of accumulated somatic mutations, increasing the genetic distinctness of individual cells and therefore decreasing the payoff that otherwise selects for cooperation. Certainly, cancer is associated with dramatic cellular mutation, and, appropriately, it is characterized by a breakdown of susceptibility to normal regulatory restraint.

In more general terms, sociobiology suggests a perspective of health. Considerations of adaptive significance suggest a functional value to phenomena that might otherwise be viewed as pathological, or, at best, neutral. An evolutionary paradigm suggests the existence of hitherto unappreciated biological strategies for fitness maximization. For example, "spontaneous" abortion may well represent one such strategy, mediated through gametic altruism, fetal altruism, and/or parental manipulation, rather than simple incapacity of fetus or mother (Bernds and Barash, in press). Similarly, the regular loss of neurons with increasing age may indicate a strategy of redundancy combined with progressive jettisoning of incompetent cells. Recent studies of the apparent value of low-grade fever (Bernheim and Kluger, 1976) reflect a similar world view, in that any consistent physiological response, such as fever in warm-blooded animals, is likely to be one that contributes to fitness, or else it would have been selected against.

A sociobiological perspective may provide numerous unexpected insights to human behavior. It holds the promise of profound insight to our innermost behavioral inclinations—those inclinations should perhaps be distinguished from the final outcomes, which derive from complex interactions between genotype and experience and which are more properly the concern of social science. For example, Dostoyevsky's ([1864] 1972) *Notes from Underground* is a profoundly disturbing work, in large part because the narrator's underlying motivation is "spite." Defined sociobiologically, *spite* refers to an interaction in which both initiator and recipient lose fitness. Accordingly, it should not be found among animals, and so far it has not been. True spiteful behavior may be unique to humans, among whom it is nonetheless generally recognized as somehow inappropriate and bizarre, perhaps because of our own unconscious application of the central theorem. Sociobiology may be more relevant to normal behavior than to psychopathology, but even here the possibility exists that further insight to neuroses and psychoses may be gained by viewing them as the adaptive consequences of individual inability to follow a behavioral course of fitness maximization. We may well have been selected to follow such courses, just as we have been selected to seek food, rest, appropriate mates, and so on.

At this point, the crystal ball grows dim. However, it seems increasingly obvious that no paradigm approaches natural selection in its ability to explain a wide range of behavioral phenomena among animals and that none offers equivalent promise of cutting a clean swath through the morass of data and theory currently surrounding research on human behavior. One thing is perfectly clear: The future of the sociobiological revolution will be neither more nor less than what we make it. When Faraday was asked about the value of his new discovery, electricity, he responded, "What good is a baby?" For good or ill, sociobiology is just such a baby.

2   *Pierre L. van den Berghe*

Bridging the Paradigms: Biology and the Social Sciences

$\mathcal{A}$fter a half century of drifting apart from the natural sciences, the social sciences are forced once more to take biology seriously. Inevitable though the rapprochement may be, it will not happen easily. The social sciences grew in the late nineteenth century out of the same intellectual movement that launched biology into its modern phase with Darwin and Mendel. Indeed, early sociology and anthropology were strongly inspired by biology and by evolutionary theory and asked all the right and important questions about human societies. Unfortunately, the biology of the time was not advanced enough to save the social sciences from giving some

As is inevitable when an established paradigm such as Darwinian evolutionary theory is suddenly being applied to a vast new body of data such as human social behavior, different minds converge on similar ideas. I therefore make no proprietary claim on any of these ideas, nor can I clearly ascribe many of them any definite parenthood. In addition to the printed sources cited, the formulations offered here owe a great deal to informal contacts, correspondence, and discussions with friends, students, of the wrong answers. Unsavory associations of early evolutionary thinking in social science with social Darwinism, racism, imperialism, and laissez faire capitalism led to a revulsion against biology among social scientists.

From the 1930s through the 1960s, the dominant current in sociology and anthropology was one of dogmatic environmentalism, extreme cultural relativism, antireductionism, and antievolutionism. There were, to be sure, countercurrents. Anthropology developed various brands of cultural evolutionism on its own. Behaviorism was not antireductionist nor culturally relativistic, but it certainly was environmentalist in the extreme and stressed ontogeny at the cost of almost complete neglect of phylogeny. As a general statement, it is no exaggeration to say that the mainstream of the social sciences in economics, political science, sociology, and anthropology was characterized by an almost complete oblivion of the organic basis of behavior.

Human behavior could be understood, it was widely believed, without any reference to the fact that humans are animals. Among the behavioral sciences, psychology and anthropology both retained an interest in human anatomy and physiology, but at the cost of internal fragmentation within these disciplines. Neither developed a consensually accepted paradigm of human behavior. Social scientists became increasingly concerned with abstract formal structures rather than with processes; with reified collectivities rather than with interacting individuals; with mentalistic constructs, ideologies, and symbolic systems rather than with observable behavior; and with statistical manipulation of aggregated data rather than with careful study of ongoing social processes. Human beings as organisms were seen as mere carnal vectors for culture and for social structures.

The era of cultural determinism in social science left a stifling intellectual legacy. During the last half century, biology has made enormous strides while the social sciences have remained

and colleagues, among whom I would like to mention Richard Alexander, David Barash, Napoleon Chagnon, Irven De Vore, Robin Fox, Penelope Greene, William Hamilton, John Hartung, Hans Kummer, Joan Lockard, Robert Lockard, Gordon Orians, Joseph Shepher, Lionel Tiger, and Edward Wilson.

largely stagnant. Indicative of this are the inordinate publication lags in social science journals: It is not uncommon for a period of eighteen months or more to elapse between submission of an article and its publication. Social scientists tolerate this because their enterprise is so uncumulative that claims for priority seldom arise. They manipulate great masses of dubious data but make few findings; they use a lot of jargon but their so-called concepts and theories are largely reiterations of old ideas, pretentious platitudes, or, worse yet, pompous nonsense. They are, however, very good at quoting classics, seeking academic ancestors, and establishing intellectual pedigrees. Their textbooks are tiresome commentaries on the gospels according to "Saints" Marx, Durkheim, Weber, and Pareto and on the epistles of "Saints" Parsons and Lévi-Strauss. During their half century of lofty isolation from the natural sciences, the social sciences have become, in short, a scholastic tradition rather than an evolving scientific discipline.

Some social scientists have given up the pretense of doing science and claim affiliation with the humanities. They should be free to pursue their worthy calling unhindered. Here, however, I shall address the majority of social scientists who continue to claim affiliation with the scientific community, and I shall suggest to them that unless their disciplines return to their biological roots their claims to scientific status are going to become increasingly tenuous. More specifically, I shall suggest that the paradigm of sociobiology, while still very much in the formative stage, is the most promising. Finally, I shall make a few suggestions on how the present chasm between biology and the social sciences can be bridged.

Before turning to that ambitious agenda, let me first briefly draw attention to the intellectual obstacles that social scientists erect between themselves and their comprehension of biological evolutionary thinking. The burden of establishing the rapprochement is squarely on the social scientists. It is they who have to rejoin the scientific fraternity. It is they who will have to throw off their self-imposed intellectual blinkers. Let me briefly examine a few of these:

Resistance to Reductionism. In the natural sciences, reductionism is widely accepted as the only game in town. In the social sciences, it is a dirty word. In their attempt to establish their *raison*

d'être as separate disciplines, the social sciences have constantly reit-
erated that human social phenomena are not reducible to any of
the "lower" levels of the organization of matter, that society is a
reality *sui generis* that can only be understood in terms of its own
laws, that collectivities are more than the sum of their members,
and so on. The high priest of the antireductionist dogma was Emile
Durkheim, but Spencer also made his contribution to it by declar-
ing society to be like an organism—indeed, a "superorganism."
Some minor currents in the social sciences, notably behaviorism,
remained reductionist, but most "schools" did not.

   Reification of the Group. Some social scientists, notably
Radcliffe-Brown and a whole school of British anthropologists,
have defined social structures as networks of individual relationships. Many, perhaps most, manage to speak of societies, cultures,
groups, organizations, social structures, norms, values, and so on
with little if any reference to individual actors. When pushed, they
will admit that societies are made up of live people, but many assume that much of what they try to understand can best be ap-
proached at the *group* level of analysis and is not reducible to individual behavior. Again, there are notable exceptions within the
social sciences, such as classical economists, game theorists, exchange theorists, and behaviorists, who make no such assumption—but even in the aggregate they do not define the field.

   Biologists, of course, overwhelmingly make the assumption
that the individual organism or even ultimately the gene is the
main unit of selection, rather than the group. There are group
selectionists in biology too (for example, Wynne-Edwards, 1962),
and biologists sometimes slip into careless, teleological-sounding
language from which one could infer group selectionist thinking.
However, the general strategy is to try to explain evolution at the
lowest possible level of organization of matter, rather than at the
highest, and this strategy has been overwhelmingly successful. It
may well be that the human species is exceptional in this respect,
but, if it is, social scientists after a century have yet to make the
kind of theoretical strides with their group level of analysis that
biologists have with individual natural selection through differential reproduction.

Dualistic Thinking. Many social scientists are afflicted with the mental malady of dichotomization. They are by no means unique in this, and possibly the human mind itself is programmed to think in terms of binary oppositions. Modern scientific thinking, however, is largely monistic in its conceptions of the universe. Concretely, the dualistic thinking of social scientists has led them to conceive of nature and nurture, of heredity and environment, as pairs of opposites. Many have implicitly assumed that, because they are dogmatic environmental determinists, biologists must be rigid genetic determinists. They have, therefore, great difficulty in understanding that natural selection theory is based on the selective effect of environmental pressures on genotypes. For biologists, the heredity-environment dichotomy is only a low-level heuristic device to disentangle causative factors in specific evolutionary processes—most assuredly not a way of stating general problems nor an issue on which to take a polemical stance. A biologist considers any stance on the relative importance of heredity and environment in the determination of human conduct as analogous to asking "What is more important in determining the nature of a coin—the head or the tail?"

Emphasis on Conscious Motivation. No doubt because humans *have* achieved a considerable measure of self-consciousness and because people often try to explain their conduct self-consciously, social scientists often assume that nearly all of human behavior must be explained in terms of conscious purpose. Some, following Durkheim, even go as far as to assume that human groups or organizations have collective consciousness and goals that transcend the individual consciousness of their members. Since biologists deal with organisms that, as far as we know, are not self-conscious, volition is not part of their conceptual arsenal; but biologists, being human, sometimes inadvertently slip into teleological and voluntaristic phraseology, as when they speak of the "purpose" of evolution. What to the biologist is a sloppy metaphor is the stock in trade of the social scientist. Perhaps the most common ground for rejecting a biological approach to human behavior is the presumably unique self-consciousness of human beings. Because humans are self-conscious organisms, it is argued, their behavior is *in prin-*

ciple not comparable to that of other animals. The statement usually stops at the level of the assertion, its demonstration being held to be superfluous because of its "self-evident" validity.

Emphasis on Verbal Behavior. The problem with an emphasis on verbal behavior is similar to the problem of consciousness. Because humans, to the best of our knowledge, are the only animals on this planet to use spontaneously, inevitably, and universally a complex communication system made up of symbols with arbitrary meanings, it is no wonder that social scientists should have become fascinated with verbal behavior and all its ideological and religious derivations. Whole brands of social psychology (symbolic interactionism, phenomenology, and ethnomethodology) come close to equating human behavior with symbolic communication. Other social scientists know that the relationship between verbal and nonverbal behavior is complex and problematic and that we do not all act as we say we do. However, verbal or written accounts of behavior are often much more easily, quickly, and inexpensively collected than rigorous observations of nonverbal behavior. Therefore, data about human behavior are typically several steps away from the actual behavior. We all too often rely on second- or third-hand reports or recollections about behavior, contaminated by selective perception, defective recall, deliberate deception, erroneous inference, ascription of motives, and many other factors that make interpretation of human behavior so difficult. We have little solid human ethology.

Emphasis on Structure at the Expense of Process. The failure of emphasizing structure at the expense of process is not unique to the social sciences, but it is characteristic of disciplines that lack a good general theory and yet seek to reduce the bewildering diversity of the world around them to a more manageable order. Biology, too, began to taxonomize on the basis of morphological structure before it could successfully explain and predict. The social sciences are still largely at the stage of description, classification, taxonomy, and empirical generalization. Since structures are often far easier to describe and classify than are processes, the structural emphasis of the social sciences is almost inevitable. Understanding process presupposes good theory, something the social sciences have only very patchily developed so far.

Interplay of Observer and Observed. The problem raised by the interplay of observer and observed also is not unique to the social sciences, but it is far more serious when observer and observed belong to the same species and are conscious of each other's actions. One of the consequences of the Heisenberg effect in social science is that a theory or a prediction has the potential of bringing forth its own negation or, conversely, of becoming a self-fulfilling prophecy. A social science theory can and often does become a political ideology used to direct political action. The Heisenberg effect thus both limits the scope of theory construction about human behavior and heightens the level of passion in scientific discourse by directly or indirectly affecting the material (and reproductive) interests of both observer and observed. The rejection of sociobiology (and of evolutionary theory in general) typically reaches fever pitch only when applied to humans. The grounds for rejection are often patently ideological rather than scientific. Modern opponents of sociobiology often think as the Victorian cleric did of Darwin's theory of natural selection: They pray that it is not true, or they hope that, if true, at least it will not become generally known.

Sociobiology is peculiarly threatening not only to social scientists but also to the generality of thinking laypersons (including some biologists) because it has an enormous potential for *demystifying* human behavior. For reasons suggested later, we evolved as a species so uniquely equipped for deceit that we have a seemingly unlimited capacity to deceive ourselves. Therefore, we can be expected to resist strenuously any attempt to strip off the multiple layers of elaborate ideological cant under which we hide our motivations.

Perhaps the most fundamental, persistent, and long-lasting debate in social science has concerned the role of beliefs versus interests in the conduct of human affairs. "Idealists," on the one hand, are those who think that humans behave in certain ways because they have come to accept norms and values taught them by their culture and that without such norms and values social existence would be unthinkable. "Materialists," on the other hand, believe that selfish interests dictate human behavior and that norms and values are either convenient conventions for everyone's benefit (such as driving on the right-hand side of the road) or the result of extraordinarily elaborate forms of deceit.

Most surprising is that the debate should continue when the overwhelming empirical evidence supports the materialist thesis. The few, limited bodies of prototheory that "work" in social science (by the usual scientific canons of parsimony, predictiveness, reproducibility, and so on) are those that are predicated on the selfish pursuit of self-interest: classical economics, behaviorism, Marxist class theory, exchange theory, and game theory. The many competing "theories"—such as symbolic interactionism, functionalism, or ethnomethodology—are not only far more recondite and less parsimonious but they are also largely unsupported by fact. The only thing that postpones their demise is that their proponents are sufficiently attuned to their own self-interests as to couch their "models" in essentially *untestable* terms. Only one social science, economics, has unequivocally accepted the selfish view of human social behavior as the net outcome of individual decisions to maximize gains and minimize losses. Not surprisingly, it is the only social science with a generally accepted theoretical paradigm, one good enough to warrant the regular employment of its practitioners by large corporate and governmental decision makers.

By sociobiology, I mean, very broadly, the application of Darwinian evolutionary theory to the behavior of animals, including humans. I am using the label that Wilson in *Sociobiology: The New Synthesis* (1975a) helped establish to describe the convergence on a general paradigm of behavior that grew out of several decades of work by behavior and population geneticists, ethologists, paleontologists, physiological psychologists, and many other specialists. But that paradigm only began to crystallize in the mid 1960s and to make an impact on the social sciences in the mid 1970s. The very label *sociobiology* has evoked passionate rejection, even among some biologists, but the simple fact remains that, well over a century after its publication, Darwin's theory remains the only viable explanation for the evolution of life on this planet. It explains behavior as well as morphology, and there is no reason to invoke special creation for our own species.

This is not the place to attempt a summary of sociobiology, for two are already available, at different levels: an easily accessible introduction for laypersons (Barash, 1977) and the lengthier and somewhat more technical work already mentioned (Wilson, 1975a). The basic theoretical literature can still be surveyed in a fairly limited number of recent publications (Alexander, 1971, 1974, 1975; Campbell, 1972; Emlen and Oring, 1977; Hamilton, 1964, 1967, 1972; Maynard Smith, 1964, 1971; Orians, 1969; Trivers, 1971, 1972, 1974; Williams, 1975). Human applications of sociobiology are, of course, by far the most controversial and are still quite tentative, but the literature on the topic is suddenly exploding (Campbell, 1975; Dawkins, 1976; Greene, in press; Hartung, 1976; Mazur, 1973; Parker, 1976; Shepher, 1971; van den Berghe, 1974; van den Berghe and Barash, 1977). The fundamental question regarding human behavior and evolution is whether self-consciousness, symbolic language, technology, and all that we mean by culture make our species so qualitatively different from all other species as to exclude us from the scope of biological evolution, at least as far as measurable change in historical times is concerned. Many social scientists have, explicitly or implicitly, taken the position that, for all practical intents and purposes, our biological heritage is of only marginal consequence in determining our behavior and that we change and adapt overwhelmingly by a cultural evolution that is Lamarckian in nature and largely operates at the level of group selection.

Sociobiology, be it noted, is *not* the antithesis of the culturally determinist position just outlined. Sociobiologists are quite happy to recognize that the human species is unique in *some* important respects. So, for that matter, is every species; otherwise it would not be a species. Humans, in short, are not unique in being unique. Nor do sociobiologists deny the importance of human consciousness and culture and the effect these have in greatly accelerating processes of human adaptation to and modification of the environment. But sociobiologists see these human attributes as the outcome of a continuing process of biological evolution and therefore as inseparable from and incomprehensible without biological evolution. There are some discontinuities but also many continuities between human and nonhuman behavior; the phylogeny of higher vertebrates clearly continues to be an essential framework for understanding human behavior. Many higher vertebrates also adapt to changing environments by social transmission of learned behavior. There is only one process of evolution. Heredity and environment, nature and nurture, the inborn and the learned—each set shows only two faces of the same interactive reality. This is as true for us as for other animals, even though our species evolved an impressive bag of tricks in dealing with our environment.

Sociobiology, then, is not a rigid genetic determinism or a simplistic instinct theory. It allows for every species evolving in its own way in meeting environmental challenges. The more complex the organism is neurologically, the more flexible its repertoire of behavioral responses and the greater its capacity for learning from its own experience. What social scientists have been saying about the conscious, purposive nature of human behavior, the complex levels of social organization made possible by symbolic communication, the high degree of environmental control made possible by modern technology, the extreme rapidity of culturally induced change, and all the other things that make us, as far as we know, unique on this planet is all undeniable. The place of social scientists as specialists in one species that happens to be uniquely dear to us is thus assured. If the social sciences are ever to achieve scientific status, however, they cannot continue to dangle in an evolutionary vacuum, isolated from the natural sciences.

Sociobiology does not challenge the separate existence of the social sciences; it merely invites them to become integrated in a theoretical framework that for over a century has been overwhelmingly successful in explaining the diversity of life on this planet. The framework is broad enough that it can easily accommodate even that argumentative animal, the social scientist.

In principle, then, there is no major difficulty in bridging the paradigm of evolutionary biology with several of the major lines of thinking in the social sciences, especially with the ones mentioned earlier: economics, exchange theory, game theory, behaviorism, and, at the "macro" level, Marxism—and even some brands of functionalism. Much of what social scientists have been saying about humans has suffered from the limitations already outlined but has nonetheless been true. Now is the time to broaden the relevance of these truths by putting us right back where we belong, as one of some two million species on this fragile little biosphere of ours. Only then will our distinctive humanity come into proper perspective.

Bridging the paradigms will be the task of a generation. Let me just make a few tentative suggestions of how this can be done. The most basic question that can be asked about human social behavior is, "Why are we social?" Some species are much more social than others, and forms of sociality also vary considerably from species to species. Among vertebrates, *Homo sapiens* is one of the more social species, but we fall far short of many invertebrate species (such as the social insects) in degree of social integration of our societies. Yet human sociality is based on more complex mechanisms and leads to more complexly differentiated societies than any other known species.

Animals band together in cooperative groups to the extent that this behavior contributes to their individual fitness (that is, reproductive success). Specifically, there are two main ways in which sociality can increase fitness: by affording protection against predators (as, for example, is clearly the case with primates and ungulates) or by affording advantages in locating, gathering and exploiting resources (mainly food), as is true of the social insects. There are also circumstances in which the clumping of resources can lead to great population densities, but mere aggregations of animals, however large, do not constitute societies unless there is *cooperative* behavior.

For primates, sociality is primarily a question of defense against predators. The problem is especially acute among the more terrestrial species, such as baboons and macaques. For hominids, predator defense may have played some role in early evolution, but it probably ceased being the *major* problem a million or more years ago. As has been repeatedly suggested, it was the ecological adaptation of early hominids to hunting and scavenging large game that required a sexual division of labor with extensive co-operation between males and females, paternal investment in provisioning the young, and intermale cooperation in hunting (Washburn and De Vore, 1961). But this was only the starting point of our sociality, for we gradually developed an ability to gang up against each other, both among closely related species of early hominids and within the ancestral line of *Homo sapiens*. We became, in short, our own predators, organized for intergroup aggression (Bigelow, 1969). Males learned to steal women, and we all learned increasingly efficient ways of wiping each other out and even of eating each other. We cooperated, in part, to kill or displace our conspecifics in competition for resources. The most successful hunting primate became both hunter and hunted, and so we remain to this day.

If such was our general evolutionary scenario, as the fossil record strongly suggests, through what specific mechanisms did we develop our complex kind of sociality? I would like to suggest that three main mechanisms evolved in succession and now continue to operate side by side to produce even the most complex of contemporary human societies: (1) kin selection, (2) reciprocity, and (3) coercion.

Kin selection. Kin selection is undoubtedly the oldest mechanism to have developed, appearing in insects that evolved hundreds of millions of years before we did. We share kin selection with all social organisms and like thousands of other social species, we are nepotistic; that is, we tend to favor kin over nonkin and to favor close kin over distant kin. We may be relatively unique in being *conscious* of our nepotism, but other animals are also nepotistic, presumably without any awareness that they are. Consciousness of kin relatedness is thus not a necessary condition of nepotism. Natural selection happens through the differential reproduction of different alleles of the same genes, which is to say that the relative success of alleles is contingent on the reproductive success of their carriers. The gene is the ultimate unit of natural selection, but each gene's reproduction is dependent on its "survival machine," the organism in which it happens to be at any given time (Dawkins, 1976).

Genes that predispose their carriers to be nepotistic will be selected for, since their duplication hinges on the reproduction of all their carriers. If biologically related individuals can be made to cooperate and thus enhance each other's fitness, then the particular alleles of the genes they share will have a competitive advantage over those carried by nonnepotistic relatives. Or, seen from the perspective of an individual organism, each individual reproduces its genes directly through its own reproduction and indirectly through the reproduction of its relatives to the extent that it shares genes with them. In simple terms, each organism may be said to have a 100 percent genetic interest in itself; a 50 percent interest in its parents, offspring, and full siblings; a 25 percent interest in half siblings, grandparents, grandchildren, uncles, aunts, nephews, and nieces; a 12.5 percent interest in first cousins, half nephews, greatgrandchildren; and so on.

If other factors are held constant, these degrees of relatedness will predict the extent of nepotism or "altruism." Relatives, to the extent that they are related, can be expected to help increase each other's fitness even at some cost to their own fitness. Theoretically, the altruistic transaction between relatives will take place if its cost-benefit ratio (to altruist and recipient respectively) is smaller than their coefficient of relatedness. In less abstract terms, nepotism is a function both of the degree of relatedness in the organisms involved and of their relative need, which in the last analysis translates into their ability to convert resources into reproduction. Parents of postreproductive age, for instance, can be expected to be more altruistic toward reproductive children than vice versa, even though their coefficient of relatedness is identical (one half in both cases).

The evidence for human nepotism is overwhelming. All human societies have been organized on the basis of kinship and marriage, and, until the rise of more complex societies in the last few thousand years, the structure of relationships created by mating and reproductive ties constituted the backbone of the social organization of all human societies. Even in the most complex industrial societies, nepotism continues to operate, along with other mechanisms of sociality to which we shall turn presently.

There is unfortunately no space here to elaborate on the application of kin selection theory to human kinship, but I have done so elsewhere (van den Berghe and Barash, 1977). Clearly, I am not suggesting a simple, mechanistic application of the model to human societies. Nor do I deny that cultural evolution has given rise to a range of elaboration and modifications on that basic biological model. Different cultures prescribe different treatment of the same category of relatives, and the same culture often prescribes different treatment of equally related kin (for example,

parallel and cross cousins). Some of these ostensibly cultural differences are themselves partly explainable in terms of kin selection
theory, because they relate to such factors as probability of paternity. The lower the probability of paternity, for example, the
more closely related matrilineal kinsmen become, compared to the
mean relatedness of patrilineal kin, real or putative. Of course, even
the simplest human societies are not organized *solely* on the basis
of kin selection.

   Reciprocity. Reciprocity has long been a central concern of
social scientists from Durkheim and Mauss to Blau and Lévi-Strauss.
Reciprocity can be said to be the basis of a relationship when
there is a conscious expectation that beneficent behavior will be
returned. It can occur between any two individuals, whether kin
related or not; but the more closely related two individuals are the
less likely reciprocity is to be a necessary and important element
of the interaction. In other words, unrelated individuals who cannot benefit by interacting on the basis of kin selection will depend
much more heavily on reciprocity for enhancing their mutual interests than will kin, who can benefit both by kin selection and
reciprocity.

   Since reciprocity is based on conscious expectation that favors will be returned, it presupposes two conditions: long-term
memory and recognition of individual members of one's interacting group. These conditions, in turn, probably exist only in the
neurologically more complex higher vertebrates (birds and mammals). The neurological capacity for reciprocity is a relatively late
development in animal sociality, and the extent to which it operates
in nonhuman species has only begun to be investigated. It seems
likely that reciprocity will be found in primates, but it is far more
developed in humans than in any other known species, and it has
probably played an important role in human sociality for hundreds
of thousands of years, certainly ever since the development of symbolic language.

   While reciprocity has been enormously elaborated on by
human cultures, it too grew out of natural selection. The prototype
of reciprocal behavior may well have been the male-female bond,
which is essentially a cooperative arrangement to raise joint offspring to the mutual benefit of both parents. It may be that only

in humans can this partnership become fully deliberate and conscious (with its elaborate ritualization in marriage), but female deception of the male to induce him to invest in offspring other than his own is not a human monopoly. Female langurs (Indian monkeys) go into false estrus when pregnant at the time a new male takes over a harem; female hamadryas baboons (a species where there is rigid polygynous pair bonding) hide from "their" males when having sneak copulation with other males. Such deceit mechanisms in nonhumans, while probably not fully conscious, give at least some indication of the evolutionary origin of reciprocity in pair mating.

The fundamental problem with reciprocity is, of course, cheating or freeloading. Assuming the universal attraction of getting something for nothing, reciprocity quickly breaks down unless there are fairly reliable ways of detecting cheaters and debarring them from further interactions. But mechanisms of detection in turn invite more subtle forms of deceit. Unless freeloaders are subtle about it, they will quickly be found out. Reciprocity, once it began to play a significant role in our hominid ancestors and to spread beyond the male-female relationship, probably became in its own right an important selective pressure for further intellectual development. Ever more subtle forms of cheating called for increasingly sophisticated methods of detection. Recent examples would be the counterfeiting of paper money or the use of computers to commit crimes. The more humans or protohumans came to rely on reciprocity in addition to kin selection as an important basis of sociality, the more it paid them to be smart. And being smart meant in good part an ability to *out*smart their partners.

Deceit, that ubiquitous feature of human interaction, probably has a fairly long evolutionary history. We clearly are highly hypocritical animals. Still, effective lying is a fine art not within the reach of everyone, since we have also evolved very subtle ways of detecting liars. The ultimate form of deception then becomes self-deception. This is where our elaborate systems of morality and ideology come in: The relentless pursuit of self-interest is usually disguised under some profession of benevolence, which is all the more effective for being sincerely believed by the speaker.

What of trust, then? Is not trust, rather than deceit, the basis of human sociality? Some would have us believe so; but what do we really mean by trust? Unless we are fools, we do *not* trust strangers. We only trust those with whom we have had good previous experience *and* those with whom we anticipate a continuing relationship. This means, in effect, that we only trust people if we are involved with them in continuing systems of reciprocity where there is a fairly reliable way of detecting, excluding, and punishing cheaters. We do not "trust" the stranger to give us a good $20 bill; we "trust" law enforcement agencies to throw counterfeiters in jail. The one major exception to the wisdom of universal mistrust is behavior toward our relatives. We often implicitly trust our close kin because we sense that, with them, kin selection reinforces reciprocity.

Even close relatives will occasionally cheat us, but they are much less likely to do so than nonkin. In fact, they can only be expected to do so when a single act of deceit is likely to bring about such overwhelming benefits as to overcome all the anticipated benefits of kin selection in the future. The faking of wills and the devious maneuvering surrounding rich relatives in their dotage would be examples of such conditions where deceit between relatives is rampant and where trust is therefore most likely to be very fragile.

Coercion. Coercion is not a human monopoly either, but there too we have pride of place. Reciprocity is nice as far as it goes, but it suffers from a gravely limiting condition: By definition, there has to be something in it for everybody. At the very least, everyone has to *think* he or she is getting something out of the exchange. Otherwise there is no point to it. Unfortunately, animals also compete for finite resources where the outcome is a zero-sum game: A's gain is B's loss. Animals regulate resource competition through some combination of dominance orders and territoriality. The former determines order of access to resources, while the latter divides the habitat into mutually exclusive patches monopolistically exploited by a single individual or subgroup within a population. Clearly, the establishment of both territories and dominance orders involves tests of strength and therefore aggression. Individual coercion between adults or of young animals by adults is quite common in many social species. Some species of primates, for example, engage in such mild forms of coercion as displacement behavior: The dominant animal displaces a subordinate for no other immediate reason than to reinforce its dominance. Others fight openly over access to key resources, such as females in estrus. A few, such as savannah baboons, even form coalitions of two or three adult males to exclude or displace other males from access to females, and a recent study suggests that reciprocity rather than kin selection may be the basis of who associates with whom in ganging up on other males (Packer, 1977).

Nevertheless, humans are coercive on a scale and in a manner quite unmatched in other animals. Human coercion (and aggression) is a group enterprise, a conscious, premeditated one. The same intelligence that enabled humans to evolve complex systems of reciprocity as a means of extending the scope of our sociality beyond the confines of kin selection also gave us the capacity to use reciprocity for purposes of coercion and thus to evolve warfare and intraspecific parasitism. Once our species had become clever enough to cooperate on the basis of reciprocity for the purpose of garnering resources such as game, the development of intergroup aggression to eliminate competitors and steal women was around the evolutionary corner. Group coercion is reciprocity for purposes of intraspecific aggression and parasitism. In a species where coalitions for reciprocal benefit can be easily formed, it is inevitable that they will be formed against conspecifics. Plunder is always tempting if I can call on enough of my partners to ensure a cheap victory over my competitors. The incidence of warfare and other forms of aggression is therefore a function of the ease with which balances of power can be disrupted. The latter can be done either through better organization or through better technology.

For all but the last few thousand years of hominid evolution, intergroup aggression was a modest, small-scale affair. Bands of hunters clashed with each other, but ecological constraints on population density were such that numbers were small and, what is more, fairly evenly balanced. Given an approximately equal level of technology and group size, the odds for success were roughly even, and therefore aggression was only moderately attractive. If, to those limiting conditions, one adds the absence of stored resources worth stealing, other than human meat and women, the attraction of plunder was also limited. One only came to blows if a particularly good opportunity for a quick kill with minimum risk of retribution arose, but such occasions were not common.

The real "take-off" point for the development of human coercion and intraspecific parasitism on a massive and organized scale was the domestication of plants and animals for food. This meant both the beginning of surplus production and much bigger population densities and thus bigger societies. With ecological constraints on societal size greatly relaxed, intersocietal competition put a premium on ever larger and more tightly organized societies, and this is where coercion took a quantum jump. No longer did loosely organized small bands simply raid each other, kill a few men, and steal a couple of women. Now much larger and better organized groups started conquering each other's land for keeps and exploiting each other's labor.

States, ruling classes, slavery, professional soldiers, courts, and all the other paraphernalia of organized coercion began to gain prominence, and soon the race was on. Ever bigger, ever more tyrannically organized societies fought ever bigger battles with each other for ever growing stakes. The rationale for internal tyranny was defense against external aggression, but the more states grew in size and power, the more coercive and parasitic their ruling classes became in relation to their own subject populations. The more internally coercive societies had a competitive edge over the stateless societies. They were, on the whole, larger and better organized for aggression. Once control over the means of organized violence became the monopoly of a small group of specialists, the machinery was, quite naturally, used for internal coercion as well as external aggression.

Conclusion

What I have done is little more than restate a well-known scenario of human evolution in a reductionist framework that stresses the following elements:

The first is a view of culture as an outgrowth of a particular line of biological evolution that is fully comprehensible only within that broader evolutionary framework. Culture is our species' way of adapting fast to changing conditions, including those of our own making. In recent millennia, it has become by far our most important way of adapting, but culture does not wipe the biological slate clean. We remain the kind of animal that our entire phylogeny has made us: a highly, but not infinitely, adaptable one.

The second is a view of human sociality as based on three main mechanisms that gained importance in clear evolutionary sequence: (1) kin selection, a blind, unconscious mechanism of gene selection through differential reproduction that we share with thousands of other species; (2) reciprocity, a more or less conscious mechanism that we may share to a limited extent with some of the more intelligent higher vertebrates but that we developed to a far greater extent than any other known form of life; and (3) coercion, a special kind of reciprocity that enables the organized few to exploit the less organized many. Coercion is reciprocity for the few at the expense of the many. All three of these mechanisms continue to underlie our sociality, although their relative weight increasingly moved from kin selection to reciprocity and from reciprocity to coercion. All three grew out of our particular line of biological evolution, but all three assume a wide range of cultural expressions.

Third is a view of collective human phenomena as the net outcome of competing individual interests. Each individual is predicted to behave consciously or unconsciously in ways that will maximize his or her gains and minimize losses. Individual interests include passing on one's genes, directly or indirectly through relatives, but in humans some interests have become at least partially divorced from reproduction. Many social scientists have asserted that collectivities have emergent properties not reducible to those of their constituent members. None has yet demonstrated that "social laws" are anything but the kind of statistical regularities such as can be expected when one aggregates any kind of individual data. For purposes of summary description, it is often handy to deal with aggregated data and to speak of collectivities as if they were independent agents. Large bureaucratic organizations in industrial societies, for example, often seem to have a quasi-organic life of their own, but closer examination reveals the complex interplay of individual interests; the supposedly collective values, norms, and goals are typically the expression of the interests of the few individuals in control.

The human being is a sufficiently special kind of animal and sufficiently dear to all of us to assure the social sciences a permanent place in our scientific establishment. But if the social sciences are to progress they must demystify human behavior. There is evidence that, because of the central role of deceit in the conduct of human affairs, we are singularly resistant to theories that demystify human behavior. Such may be the main basis of opposition to reintegrating the social sciences into a theoretical framework that accounts for the evolution of all life forms on this planet. But demystify we must if we are to reach an adequate understanding of our own behavior. Given our rapidly growing ability to destroy ourselves and our habitat, it is urgent that we improve our level of self-consciousness about our behavior.

Our dilemma is that if we continue to behave as we have in the past we shall greatly hasten our inevitable disappearance from this planet. In the long run, as John Maynard Keynes remarked, we are all dead; but as far as our species is concerned behavioral change can mean the difference between five hundred years and five billion years. Sociobiology predicts that we shall continue to reproduce, consume resources, and destroy each other with abandon because we are programmed to care only about ourselves and our relatives. So far, there is little evidence to show that sociobiology is wrong. The ultimate challenge of humanity is to prove sociobiology wrong, not by assertion but through self-conscious change in our behavior. Far from being an apology for the status quo, sociobiology is a challenge for change. The more we learn about the kind of animal we are, the more self-conscious our behavior will become; and the more self-conscious we are, the more effectively we can change in the direction we choose.

3

S. L. Washburn

Animal Behavior and Social Anthropology

*I*nterest in animal behavior has increased sharply, as shown by the number of programs on television, the success of popular books, and the recent appearance of reviews and textbooks. This general interest has been further stimulated and channeled by Wilson's *Sociobiology: The New Synthesis* (1975a), which calls for a new synthetic science of animal behavior; the topic has immediately become popular and has taken on many of the qualities of an intellectual movement. Its adherents are enthusiastic, and they do not welcome criticism.

From the point of view of social anthropology, I think that it is essential to keep the very heterogeneous study of animal behavior separate from Wilson's (1975a) version of sociobiology. Clearly, the appeal of sociobiology is that it offers an evolutionary

The program on the study of human evolution has been supported by grants from the L. S. B. Leakey Foundation and the Wenner-Gren Foundation for Anthropological Research. I wish to thank these foundations, Alice Davis for editorial assistance, and Burton Benedict and Robert K. Colwell for criticism and helpful suggestions.

and genetic explanation for a remarkable diversity of animal behaviors, but equally clearly the theory leads to gross errors when
applied to human behavior. The reasons for this are complex, and
I can see only a few of them. What I will try to do in this chapter
is to suggest that social anthropologists may gain many insights into
human behaviors by considering the behavior of other animals, but
sociobiology (as practiced at present) will only bring confusion to
anthropological problems.

   As I read anthropology, I see a long history of biology confusing and retarding the development of the social sciences. Nineteenth-century evolutionism, orthogenesis, reductionism, biological analogies, homeostasis, racism, IQism, eugenics, and the ever
present confusion of genetic and environmental causes—all have
been major liabilities to social science. History cannot be changed,
but a strong case can be made that the founders of social science
would have been far better off if they had never heard of biology
or evolution. But at the same time that biology was a liability to the
social sciences major progress was being made in the biological
understanding of human behaviors.

   At present, a great deal is known about the biology of our
species, information essential for the practice of medicine. Information on a very wide variety of behavioral topics (sleeping,
dreaming, thinking, speaking, to mention only a few) is available
for social scientists. But the use of this kind of knowledge in trying
to understand human biology in no way contradicts Durkheim and
the idea that social facts must be explained by social facts. The attempt to understand the biology of the human actors, of the social
systems, and of their interrelations are three supplementary kinds
of information. Problems arise when the differences between the
differing kinds of information are blurred. Sociobiologists must
attack Durkheim and the logic of social facts, minimize the distinction of heredity and environment, and postulate genes to account for behaviors—a repetition of the errors of previous biological confusions.

   The fundamental way sociobiology creates confusion may
be illustrated by a quotation from Hamilton (1975). Hamilton's
papers on genetics form the foundation of sociobiological thinking

and are referred to in all papers on sociobiology. The following paragraph illustrates what may happen when a person who has made major contributions to the theory of natural selection (inclusive fitness) discusses human evolution. The following nonsense (Hamilton, 1975, pp. 149–150) belongs in the era of laissez faire capitalism and Spencerian social evolution.

> The incursions of barbaric pastoralists seem to do civilizations less harm in the long run than one might expect. Indeed, two dark ages and renaissances in Europe suggest a recurring pattern in which a renaissance follows an incursion by about 800 years. It may even be suggested that certain genes or traditions of the pastoralists revitalize the conquered people with an ingredient of progress which tends to die out in a large panmictic population for the reasons already discussed. I have in mind altruism itself or the part of the altruism which is perhaps better described as self-sacrificial daring. By the time of the renaissance, it may be that the mixing of genes and cultures (or of cultures alone, if these are the only vehicles, which I doubt) has continued long enough to bring the old mercantile thoughtfulness and the infused daring into conjunction in a few individuals who then find courage for all kinds of inventive innovation against the resistance of established thought and practice. Often, however, the cost in fitness of such altruism and sublimated pugnacity to the individuals concerned is by no means metaphorical, and the benefits to fitness, such as they are, go to a mass of individuals whose genetic correlation with the innovator must be slight indeed. Thus civilization probably slowly reduces its altruism of all kinds, including the kinds needed for cultural creativity.

This absurdity has been highly praised by Fox (1975, p. 7), and this kind of writing does not seem to disturb sociobiologists.

Obviously, the importance of this particular passage lies in the kind of thinking it represents. It precisely repeats the errors of the early evolutionists who thought that their theory was so powerful that facts could just be arranged in order *without* doing the necessary research. Stone tools were arranged in evolutionary orders without archeological information. The orders proved to be wrong. The mistaken notion that evolution necessarily moves from the simple to the complex led to countless hypothetical reconstructions. Close to a hundred years of mistakes were justified by the belief in the power of the theory of evolution. Now, the general theory of evolution is correct, but a theory does not give conclusions—it directs the nature of the research, but each application of the theory demands careful research.

This is the fundamental weakness in the passage by Hamilton just cited and in the majority of sociobiological thinking as applied to human behavior. Writers are so confident in the power of the theory (selection, adaptation, inclusive fitness) that a minimum effort is made to learn the facts of human behavior and of human history.

In analyzing the whole Hamilton paper, I find that: (1) Useful genetic theory is briefly presented and clearly discussed. (2) Practically no effort is made to uncover the facts of human evolution, of recent human history, or of animal behavior. (3) The conclusions are personal biases, which stem from neither facts nor theories.

This is the general form of the application of sociobiology to the interpretation of human behavior. It is for this reason that it is repeating the mistakes of many years ago. There is no way for a scientist to leap directly from genetic or evolutionary theory to conclusions about human behavior. The principal task for the scientist is the research that links theory and conclusion. Lacking this link, sociobiology may be useful and illuminating, or it may be reductionist, racist, and absurd.

For example, although sociobiologists speak of human evolution, practically no attention is paid to the archeological record. The record shows that, starting a little before agriculture, there was a very rapid change in technology that spread all over the world in a very few thousands of years. The conditions that had dominated human evolution for some millions of years changed; the nature of the change, the speed of the change, and the rate of diffusion all show that the changes were the result of learning, not of biological evolution. Inclusive fitness theory states that an individual's social actions should lead to the perpetuation of the person's genes, either through descendants or through the survival of relatives. But in the quotation from Hamilton it should be noted that, with civilization, benefits to fitness will no longer go to closely related individuals. In like manner, worried by increased mobility, Eshel (1972) thinks that evolution may no longer favor altruistic genes. To generalize the problem, the more people there are, the more mobility, the more variety of selective pressures, and the shorter the time span—the less will inclusive fitness theory be useful. And what the archeological record shows is a great increase in all these factors, beginning with late Paleolithic and Mesolithic times and accelerating to the present. Today an individual's interactions are mostly with nonrelatives, and millions of people are deliberately reducing the number of their offspring. The conditions that dominated the evolution of *Homo sapiens* no longer exist, and a theory of evolution cannot be indiscriminately applied to human evolution as if the contemporary situation were similar to that of thousands of years ago.

The very rapid increase in population in the last 200 years is certainly the result of technology, not of changing gene frequencies. As earlier and earlier times are considered, it becomes increasingly difficult to prove that the basic changes were due entirely to learning, not genes. My belief is that there has been no important change in human abilities in the last 30,000 years. If this is even in large part the case, biological evolution cannot account for the social conditions of the modern world. This is certainly not a new idea, but it is the historical reason for stressing the importance of learned social facts rather than genes.

It has become fashionable to minimize the nature-nurture argument by stressing that both are important, that animal behavior is more plastic than it had been previously considered, and that human behavior may be more determined by genes. For example, Emlen (1976) suggests that the importance of the distinction has been exaggerated. Sociobiologists must take this position because they want to emphasize the importance of genetics in accounting for human social behaviors (again, note the quotation from Hamilton). As far as the interpretation of human social behavior is concerned, social anthropology will regress at least fifty years by allowing the distinction to become blurred. Given populations of *Homo sapiens,* social behavioral differences are to be understood as the results of history, of different groups of human beings learning different languages and ways of life.

The essential confusion in the comparison of human behavior with that of other animals is in the nature of learning. Confusion at this level lies behind most of the problems of comparing human social behaviors and will constantly appear and reappear in a wide variety of forms when biologists and social anthropologists discuss behavior.

In these discussions, the term *culture* is not too helpful a word, partially because there are so many definitions and usages and partially because there is more learning in animal behavior than had been initially believed. Therefore, for the purposes of this chapter it is useful to regard language and all the complexity of behaviors that language makes possible as the basic difference between human and nonhuman. Language—the complex of brain, speaking with a phonetic code, and perception of verbal patterns, unique to human beings, a product of singularly human evolution—is a behavior not found in any other animal. Language is the ability that makes the nature-nurture problem in the human species special and almost totally different from the comparable problem in other species. Without language, human behavior might be interpreted along rules similar to those governing the behavior of other animals. With language, the rules change, and human social behaviors cease to be under genetic control.

Human language is basic to environmental reference and social reference. Technology, economics, social relations, political systems, religions, arts—all human activities—depend on language, and it is language that makes it possible for humans to learn complex systems of behaviors. It is my belief that only very minimal technical progress is possible without at least some simple form of language, and, of course, no complex technology or science would be possible without this special form of communication. From an evolutionary point of view, it is not only that the human brain has evolved but also that the brain can easily learn to manage quantities and varieties of information without parallel in the animal world. The communication of this information depends on language. When considering human behaviors, *it is language that gives the nature-nurture problem a major new dimension, vastly increasing the realities and possibilities of learning.* When comparing human and nonhuman behaviors, any theory that minimizes the importance of distinguishing between linguistically mediated learned behaviors and other behaviors, in which learning may be much less important, is bound to cause confusion.

The Comparative Study of Behavior

The comparative study of behavior has been a very important element in the rise of biosocial science. But there is no agreement on how comparisons are to be made. For example, a recent book on the biological bases of human social behavior includes nothing on the brain, little on hormones, and a scrap (misleading) on language; the author apparently considers the study of behavior to mean looking at the outsides of animals. Comparisons frequently take a pseudo-evolutionary form, although this method has been criticized for many years (Hodos, 1970; Hodos and Campbell, 1969), and the general sort of comparison usually made cannot possibly deal adequately with human behavior. Comparisons of animal behavior have been used in many different ways, and there must be a clarification of the uses of the comparisons.

Comparisons Are Fun

Quite aside from any practical benefits, it is interesting to learn of the language of the bees, pheromones, factors guiding migrations in fish or birds, the life of the penguin, or the world of the herring gull. Just as the discovery of the gorilla or okapi stirred former generations, so the analysis of animal behavior excites many today. The success of myriad television programs and popular books shows the interest in behavior, and topics such as ant navigation by polarized light demonstrate a very sophisticated science, a far cry from traditional natural history. If one aim of education is to learn about the world, then an appreciation of the world should include some understanding of the diversity of life and the extraordinary variety of behaviors. Typically, in this kind of comparison, enormously diversified groups of vertebrates—even insects—are compared to a single group of human beings, and no attempt is made to show that there is a common biological basis for the behaviors being compared.

I like Chance's notion that comparisons may alert us to new thoughts, new points of view. Attention structure or critical periods suggest ways of looking at human behaviors. Experiments on monkeys direct attention to the possible importance of early experience in humans. We are alerted to the existence of whole new sensory

worlds and social systems that work with almost mechanical precision. But neither comparisons nor alerting are proof. They offer
suggestions, and a very different sort of information is needed if
the information is to be seriously used in the solution of human
problems.

Serious Comparisons

It is natural to want to apply the understandings that come
from animal behavior to the behavior of human animals. To do
this, it is crucial to decide when such comparisons are useful and
when they are not. For example, Lorenz's *King Solomon's Ring*
(1952)—a great book, especially considering when it was written—
belongs to the "fun" stage in the development of animal behavior.
It interested many people in animal behavior and was instrumental
in the establishment of ethology as a distinctive science. *On Aggression* (Lorenz, 1963) is not a helpful book, mainly because it uses
bits of behaviors from oddly selected animals to suggest causes for
important human social problems. It closes with advice that does
not come from the data and can only be characterized as irrational:
"the main function of sport today lies in the cathartic discharge of
aggressive urge" (p. 280).

Lorenz has been criticized for *On Aggression,* but the issue
is not the author; it is the method. For example, in one argument
Alexander (1975) mixes information about insects, birds, and Ti-
betan peasants—a discussion meaningless in serious comparison.
The final chapter in Wilson (1975a) is an extreme example of the
misuse of comparative information. The point can be illustrated
by comparing Wilson's treatment of insect behavior with his treatment of human behavior. When discussing insects (where he has
facts), he does not bring human beings into the discussion. But
when discussing human beings (where facts are minimal), insects
are used to make critical points. And, in passing, it might be
pointed out that, in spite of countless references to the contrary,
the castes of India are not genetically determined and bear no resemblance to castes of insects.

Serious comparisons require careful, detailed analysis and
experiments whenever possible. For example, the facial expres-

sions of humans and monkeys have often been compared. But massive removal of the cortex of the brain in monkeys affects facial expression minimally (Myers, 1976); comparable damage in humans would result in total facial paralysis. The experiment shows that human facial expression is far more under cortical control, far more amenable to learning, than is the case in nonhuman primates. In monkeys, both sounds and facial expression are primarily under control of the primitive emotion-controlling parts of the brain (limbic system). Just comparing monkey and human expressions and sounds cannot reveal the major structural differences underlying both facial expression and sound control.

In comparing the biological bases for human social activities (such as being social, speaking, gesturing, and feeling various emotions), differences in the brain are always involved, and comparisons are necessarily complex. For example, the human family is a remarkably varied institution, serves many functions, and cannot be usefully described in any simple way. But, if the family is described as a "pair bond" (Larsen, 1976), then all the variety stemming from the uniquely human brain, language, history, and complex function is lost. One may then look for other creatures that pair bond and end up with the notion that the human family may be better understood by studying herring gulls. This example illustrates two common mistakes in the comparative study of behavior. First, any chance to understand the human condition is lost by reducing the understanding of the family to the expression "pair bond." Second, it is a useless exercise, because pair bonding, where it does occur, can only be understood by analysis, not by uncontrolled comparisons. Gibbons do pair bond, but to understand this behavior it is necessary to appreciate a way of life in which locomotion, diet, and territorial behavior are all interrelated (Washburn, Hamburg, and Bishop, 1974). Lorenz points out that comparisons between closely related forms are the most likely to be useful. This wise advice is almost never followed when human behaviors are compared with those of other animals.

Ethology has been so important in the rise of animal behavior that it merits special attention. To my mind, the genius of ethology has been in its employment of experiments, not in the study of natural behavior, as is well illustrated in Tinbergen's classic *The Herring Gull's World* (1960). Interpretations were constantly checked by experiments, and the world of this bird is revealed as being something very different from what it looked like to the observing human. The results of ethological investigation are fascinating, and, while *Ethology: The Biology of Behavior* (Eibl-Eibesfeldt, 1975) is well worth careful consideration, the chapter on the ethology of humans shows how misleading the ethological approach can be. By investigating human behavior with the questions and techniques suitable for animals with very simple nervous systems, the whole nature of human behavior is lost. Human behavior cannot be understood by observation alone, and, even when considering infants who cannot yet talk, the mother is part of a way of life based on speech, complex social behavior, and technology. The notion of cataloguing behavior in an ethogram or biogram comes from the study of the behavior of a few animals—largely in captivity— and is completely inappropriate when applied to humans. In terms of anthropological history, biograms and ethograms belong with prefunctional anthropology, and functioning systems do not lend themselves to that kind of cataloguing. Whether the concern is biological mechanisms, biosocial functional complexes, or the elaboration of symbolic systems, the notion of the ethogram (behavior catalogue) is misleading and provides no useful guide to research.

From a practical point of view, the easiest way to avoid gross reductionism is to start comparisons with human behavior. Our heritage from evolutionary thinking leads to almost all comparisons being made in the opposite direction. The problem ethologists have with human social behavior is that in their whole strategy of research there is no place for the critical importance of language and learning nor for the biology of cognition and language as being the facilitating mechanisms for uniquely human behaviors. Human ethology might be defined as the science that plays a game: It ignores the fact that humans can speak. No one who starts with an understanding of human behavior will speak of pair bonds or compare the complex social and technical hunting patterns of human beings with those of the biologically based hunting patterns of carnivores (unaided by weapons, performed by both sexes, sometimes at speeds of more than twice human Olympic records).

If the "fun" kind of comparisons (aggression in fish, birds, and Yanomamö Indians of Brazil) are replaced with serious comparisons, then study reveals differences rather than similarities and reveals that uniquely human behavior is based on and facilitated by uniquely human biology. In summary, animal behavior is a fascinating study in its own right. It is pleasurable and basically liberating in the best traditions of a free and basic science. But at the present time most comparisons are far more likely to be misleading than helpful to social anthropologists.

The methods of ethology have been defended as being similar to those of comparative anatomy (Blurton-Jones, 1975, p. 70). But the methods referred to are the very part of comparative anatomy that has worked only to a very limited extent or in some instances not at all. Specifically in the case of humans, no one has reconstructed a fossil ancestor that looked like *Australopithecus* prior to the discovery of the actual bones. The lesson from comparative anatomy is that generally there must be fossils to prove an actual evolutionary sequence. Even in anatomy, there is no way to triangulate back from present forms to reconstruct ancestral populations, and the case is even worse for behavior.

The problems of evolutionary theory and social anthropology may be illustrated by the conclusions of some of the leaders in the field of evolutionary biology. Alexander (1975) concludes that society is based on lies. Hamilton (1975) thinks infusions of pastoral genes are necessary to keep creativity. Wilson (1975a) thinks neurobiology is necessary for ethics. In addition to technical disagreements, especially over group selection, these authors read like a mixture of Herbert Spencer and science fiction.

Various reasons lie behind the confusions of evolutionary theory when it is applied to human beings. Perhaps the most fundamental is the belief that if the theory is correct the application of the theory is easy and routine. But some general theories—the notion of selection leading to adaptation, for one—are only guides for research. If human locomotion is the area of interest, then the structure of the pelvis and the lower limbs must be interpreted in terms of the way they function. This adaptation is unique to humans, and there is no way that this form can be predicted by the theory. Nor is the form resulting from the evolutionary process necessarily efficient or ideal. It is just what, as a part of a complex process, succeeded in competition with others. In the female, adaptations for locomotion and easy childbirth are in conflict. In both sexes, the lower back is too weak. The ligaments of the knees should be stronger. The mechanics of the ankle and foot could easily be improved. Perhaps in a few million years the adaptation might have become more efficient, but in the short run the defects may have to be remedied by the surgeon. Even in behaviors that are largely genetically determined, there is no reason to suppose that the behaviors are necessary, ideal, or more than the compromise that survived in evolutionary competition.

The application of the general theory of evolution (selection, adaptation, inclusive fitness) requires that the facts used in fitting the theory be correct, and the theory itself only suggests where to look for facts and which ones may be relevant. This is the fundamental weakness of the authors I have cited. They know theory, but they know very little about human behavior or the few known facts of evolution. The power of the theory gives them confidence that their personal opinions are correct and should be adopted.

The theory that biology is adaptive is very similar to Malinowski's (1944) theory of functionalism. Both are useful as guides to research, and I have often pointed out that Malinowski's *A Scientific Theory of Culture* (1944) is a better guide to the behavior of the nonhuman primates than to language-based human behavior! But in neither biology nor human behavior may one simply assume that the fact one is dealing with is adaptive—or even a fact. Consider the nose, which is obviously adapted for respiration. Its function is to admit and to warm air, so a great many papers have been written attempting to correlate the external form of the bony nose with climate and temperature. But the nose is also the middle of the face, and the incisor teeth form in the bone just below the nasal opening. As can be seen in growth patterns or by comparison with other primates, the nasal width is in part a function of the size of the teeth, and the center of the face must be seen as a complex involving several factors.

One may not look at a structure or a behavior and simply assume an adaptation. For example, Alexander (1974) postulates that the menopause may have evolved because women reach an age at which it is advantageous for them to take care of the already present offspring rather than to have more—suggesting positive selection for menopause, a genetic explanation. But an alternative theory is that the increasing efficiency of the human way of life (culture) made it possible for people to live longer and longer. With no genetic change, women would live past their reproductive period, a process still going on. There is no doubt that the length of human life has increased greatly over the last century without genetic change. One explanation is highly speculative and genetic; the other is environmental and can, at least in part, be directly verified.

Postulating positive selection of genes for menopause blocks consideration of the increase in length of life. Sociobiologists must postulate genes for behaviors because they would not have a science if they did not do so. Postulating genes for behaviors has been so prevalent in European thinking (eugenics, racist theories, genes for crime, genes for historical change—see the quotation from Hamilton at the beginning of this chapter) that it deserves a special name. I suggest *genitis* ("geneitis," the genetic disease). This disease consists of postulating genes to account for behaviors without making any major effort to see if the suggestions are reasonable. Most sociobiologists seem to have bad cases of genitis. Sociobiologists postulate genes for altruism and others for cheating, but it would be far more adaptive to possess genes for intelligence and be able to cheat or be altruistic as occasion demanded. Genitis leads to futile, uncontrolled speculation.

Evolution and Semantics

The basic problem with postulating genes for social behaviors is that it shifts the nature of explanation from the logic of social facts to the logic of genetics. Even though the environment may be important in both cases, it makes a great deal of practical difference if the cause of crime is in large part genetic or if *crime* is a word standing for a wide variety of behaviors learned under varying circumstances and defined differently in different cultures. At the present time, the word *evolution* has no clearly defined meaning, and it may be used in a way to confuse biological and social explanations. For example, in his presidential address to the American Psychological Association, Campbell (1975) defined both biological and social evolution as "blind variation and systematic selective retention." But Lenski and Lenski (1978, p. 79) define sociocultural evolution as "technical advance and its consequences." The best evidence for sociocultural evolution comes from the part of human behavior that is due to purposeful technical progress and that is least appropriately labeled "blind." Naroll and Divale (1976, p. 97) define evolution as moving "in the direction of increasing functional complexity" and state that this is the case in biology, giving Simpson as a reference. But Simpson (1949, p. 253) stated that "evolution has . . . often and . . . misleadingly been generalized as just a succession from simple to more complex forms of life." Evolution from complex to simple is not just associated with loss (as in cave fish), but the human skull is also far simpler than the skull of the ancestral fishes, and the human brain is more complex. The ordering of human historical and social facts cannot be justified by appeals to biological evolution, nor can the use of the word *blind* do more than confuse genetic change with social change.

I have used recent references because these old problems are still very much with us. If it is desired to use words that stand for similar operations, then the following words have the same meaning—*evolution, biological evolution,* and *biological history*; and then the methods of study in these topics are genetic and biological; and they assume that change requires a long period of time. Another series of comparable words are *evolution, social evolution, sociocultural evolution,* and *history*—for which the methods of study are concerned with the record of learned human behaviors. The two kinds of evolution may be in a feedback relation, but the processes involved, methods of study, and implications for social science are radically different. It is my belief that when people resort to the word *evolution,* even when qualified by *social* or *sociocultural,* they believe that something has been said that is more important than referring to the same chronological events as *history. Evolution* is a magic word (Washburn, 1976), and it is used to predispose the reader to certain kinds of conclusions. However that may be, to avoid confusion it is necessary to consider evolution (genetic change) and history (cultural change) separately before the interrelations between the two are considered.

Evolution

If there are problems in the comparison of behaviors, these are compounded in the study of evolution. To the difficulties of comparison are added the complications of the fossil record, time, and constantly changing theories. Obviously, the challenge is that—with enough facts and the right theories—evolution, and *only* evolution, will give the overall synthesis. Evolution provides the master view of what happened and why it happened. But in spite of the technical power and intellectual magnificence of the evolutionary perspective, it is by no means clear how it can be applied in ways helpful to social anthropology. As noted earlier, there is a long history of biology hampering progress in social analysis, and, as far as social anthropology is concerned, it is wise to distrust the claims of evolutionary biology. Clearly, the problem is that both biology and social anthropology are changing, and, from the social anthropological point of view, the question is whether or not people are more likely to be effective social anthropologists if they know some biology.

Just as in the case of comparisons, the study of human evolution must start with facts about human beings. Only by studying human beings can we know that they are bipeds and have large brains, great cognitive abilities, language, religion, arts, and a complex social life. There is almost infinitely more information about humans than there is about monkeys or apes. The problems cannot be usefully framed by starting with animal behavior, particularly the behavior of forms in which behavior is largely genetically determined and stereotyped. Most human behavior has no close counterpart in any nonhuman animal, and this is the most important conclusion of the study of human evolution. For this reason, the application of general theories, which apply to very different forms of life, must be applied to humans only with the greatest caution.

For example, the nature-nurture dichotomy was unfortunate; it is a fact that most behaviors have genetic bases modified by environment. But the relative contribution of genes and environmental influence may vary from almost nothing to 100 percent. There is no evidence that there is any genetic component determining the differences in human languages or social systems. Biology forms a facilitating base, and the importance of understanding that base has been stressed earlier, but—at the risk of repeating—I again stress that these negative evaluations of the comparative study of animal behavior and of recent applications of evolution to the final stages of human history are only negative in the context of the importance of these studies for social anthropology. In my opinion, they have very little to offer, and the conclusions are usually misleading. The same evaluations probably hold for social science in general. But the study of animal behavior is fascinating and should be a part of everyone's general education. It opens a new view of the world, a view very different from traditional beliefs. Animal behavior, as described in such books as those by Alcock (1975) and Brown (1975), is as different from folk belief as modern medicine is from folk medicine. Evolution is a master theory, accounting for the origin and diversity of life. It is the great materialistic theory that takes the place of the idea of creation and helps in the understanding of the biological nature of the world and of human beings. But the existence of varying societies and cultures is made possible by human brains, language, and learning, and the rules of social learning are fundamentally different from those of evolutionary genetics.

Genetics

Sociobiology is partially the result of a renewal of interest in natural selection, evolution, and adaptation. To the traditional formulation of these problems has been added the concept of kin selection. Barash (1977, p. 85) gives a very clear account of the history of the concept and its consequences. According to traditional selection theory, there was no way of accounting for acts that appeared to reduce the fitness of the individual performing the acts (altruistic acts). Kin selection shows that there may be selection for altruistic acts, if they benefit the relatives of the altruist. Relationship is measured by the coefficient of relationship, which gives a measure of the proportion of genes that two individuals share because of common descent. For identical twins, the coefficient will be 1. For entirely unrelated individuals, the coefficient will be 0. Between parent and offspring, the coefficient is ½; between full siblings, ½; between aunts and uncles and nieces and nephews, ¼; between cousins, ⅛.

According to sociobiologists, social structure should follow this genetic calculus. The patterns of actions between human beings should follow the same pattern as the genetic relations. Sahlins (1976b) has shown that this is not the case, but it is too soon to tell whether his analysis will exert any influence on sociobiology. In modern mobile societies, most interactions will be with unrelated individuals (coefficient 0), and this is why Hamilton (quotation cited earlier—1975, pp. 149–150) thinks that altruism and creativity must be reduced in civilizations, although all the historic evidence suggests that precisely the opposite is the case.

There is a fundamental complication in the analysis of kin selection. Although an individual receives half his genes from each parent, an individual shares half the genes in which the parents differ and all the genes shared by both parents. Within a species, the majority of genes are shared, and it has been estimated that two human beings selected at random will share something on the order of 90 percent of their genes. The similarity is much greater if it is based on estimates of the structure of DNA of the two individuals. For example, King and Wilson (1975) estimate that human beings and chimpanzees share 99 percent of their genetic material and that human races are fifty times closer to each other than humans are to chimpanzees. This should not be surprising if the anatomical similarities shared by all human beings are considered. But it means that individuals who would be considered unrelated (coefficient 0, as normally calculated) share almost all their genetic material.

There are three consequences of stressing shared genes and common biology, as opposed to stressing differences. First, traditional genetics could only find differences. If there were no contrasting allele, a gene could not be defined. To a large extent, traditional genetics had to be a science of difference, often of trivial difference. But with the understanding of DNA came the possibility of seeing similarity in the genetic substance, and the study of DNA shows that there is an enormous amount of duplication. It is no accident that sociobiologists pay no attention to DNA or species-wide similarities that determine behaviors.

Second, if behaviors that are important to the basic adaptations and survival of the species are based in large part (not completely) on homozygous genes, then as far as these adaptations are concerned random mating will have close to the same results as the mating of close relatives.

Third, if in the case of human beings we are concerned with the behaviors that form the base of the success of the species (bipedalism, hand skill, cognition, language, ease of learning, and social complexity), these are all highly complex and, with rare exceptions, common to the whole species. Therefore, there is no reason to suppose that customs favoring close relatives would have great effects on the evolution of the biology that is important to human social behaviors.

One of the fundamental controversies in genetics is group selection, and this is of particular importance to social anthropologists. In general, sociobiologists deny the importance of group selection, but it is strongly supported by Wilson (1976b), and Alexander (1974, p. 376) states that "human social groups represent an almost ideal model for potent selection at the group level." What Alexander means is what anthropologists would call *culture* and what in this chapter I have viewed as the results of language. Shared social knowledge may be adaptive, but not because of genes.

The issues may be simply stated as follows. It is individuals who leave offspring. If the lives or deaths of individuals change gene frequencies, then there is evolutionary change. The gene frequencies in populations influence the phenotypes in the next generation and are affected by the differential survival of individuals. But the survival of individuals may be determined by the group in which they live. The more knowledge, technology, and social organization (culture), the more important the group is in determining survival. We have got the Gatling gun and they have not. The groups form a species, which is in competition with other species.

If evolution is defined as the changing gene frequencies in populations (Dobzhansky, 1962), if the changes are due to differential survival, and if the survival may be strongly influenced by the group, then some consequences follow. Contrary to much sociobiology, war may not have any effect on evolution at all. It is only if the gene frequencies of survivors are different from the vanquished that there is any effect on evolution, on changing the frequencies of genes. If most wars through most of human history have been with close neighbors, the effect on evolution has probably been minimal. Further, it makes no difference if some people have many offspring and some none, unless this changes the gene frequencies of populations. For example, if celibate monks (taken as a group) and the males leaving most offspring (taken as a group) have the same gene frequencies, the evolutionary process is unaffected. There may be social change, demographic change, but evolutionary change can only be demonstrated if it can be shown that differential reproduction was accompanied by genetic change.

In summary, there is no clearly defined, universally accepted evolutionary theory that social anthropologists must accept. There has been great progress in the understanding of genetic mechanisms, but naturally there still are fundamental controversies. To my mind, the most fundamental problem comes from postulating genes to account for behaviors.

Biology and Human Behavior

At the present time, I think that the evolutionary, comparative, and sociobiological approaches to human social behavior are far more likely to cause confusion than to be helpful. As practiced today, these approaches repeat long-standing errors and perpetuate a way of thinking that took shape more than a century ago. Perhaps anthropologists in particular are caught in a mode of thinking of which they are hardly aware—thus the purpose of these last few pages is to suggest that substantial change is needed.

The general issues may be made clear by two examples. From the point of view of behavior, sleeping is a fundamental problem, although social scientists pay very little attention to it. From the evolutionary point of view, there is no direct evidence on sleep. If sleep is viewed as a part of the problem of circadian rhythms, then there is no direct evidence on this whole class of problems, and the patterns of sleep of our distant ancestors remain a matter of opinion. From the comparison point of view, there is fascinating information on sleep behavior in many kinds of vertebrates (Snyder, 1966). The complex problems of timing of sleep, locations for sleep, and avoiding predators during sleep form a set of problems with a wide variety of behavioral solutions. But the comparative behavioral information is literally superficial, and it does not help in understanding the underlying biological mechanisms. The mechanisms are revealed by experimental intervention in animals and clinical study of human beings, and the comparative study of behavior alone could not even reveal that there is a reticular formation nor lead to understanding of its activating function.

Human beings have always been concerned with the problem of dreams, on which there has been a great deal of speculation in psychology and anthropology. Dement (1974) has written a readable account of the modern, experimental study of sleep and dreams, and any social scientist with interest in this area would be far more likely to make useful contributions starting with what is known scientifically about sleep, rather than with folklore.

The same general point can be made with regard to sex. Anthropologists inevitably consider sexual customs, the recording and analysis of which are far more likely to be useful if based on modern biological understanding (Katchadourian and Lunde, 1972). It is easy to forget how recent such understanding is. The interpretation of the female cycle depends on experiments performed in the 1920s (Corner, 1942), and earlier confusion between menstruation and bleeding in nonprimates shows how limited comparisons may be if unaccompanied by experiments.

Generalizing from these examples, the point is that human beings are more likely to behave reasonably if knowledgeable than if ignorant. Since many human decisions involve human biological behavior in one form or another, everyone should have some basic biological knowledge. This is especially the case for those who would interpret human behaviors. So, if one is interested in human behavior (social or biological), one should start with what is now known about human beings. Then one can move to comparison or evolution with far less danger of error than if the process is reversed.

Putting the matter in a more general form, human biological individuals are the actors in the social systems that social scientists describe, and the biology of the actors is in immediate and complex relations with the social behaviors. Durkheim's social facts are in the minds of biological beings, and implicit in the treatment of social facts is that the persons in the social system are normal members of the species *Homo sapiens.* Considering the actors means that the social rules only affect any actual individual through the behaviors of other human beings, and this removes any suggestion that the rules will be automatically followed, as well as any suggestion of the superorganic.

No social system may be fully described or understood without considering the biology of the actors. This simple statement has profound consequences for the explanation of any social system.

First, it is not an analogy. Human actors are necessary for human social systems, and termites are necessary for termite systems.

Second, it is not reductionist. The behaviors of the actors may be found to be simple or complex, largely based on learning or largely genetically determined. The interrelations of behavioral biology and learning are empirical questions.

Third, no claim is made that biology is important in all questions.

Many of the present worldwide problems are biosocial. Population, nutrition, health, education, and interacting in satisfying ways—all these have biological, social, and historical factors and often interact in unexpected ways.

When the American school system places boys and girls in classes according to chronological age, ten-year-old girls consistently outperform ten-year-old boys on many tasks, especially spelling. But an important reason for this is that the girls are biologically much more mature than the boys (by some eighteen months). The biological facts clearly affect the social situation, and the practical situation may be analyzed without reference to comparisons or evolution. However, I think that adding these dimensions may give additional insights. The function of the delay of maturation in human beings may have been primarily the result of the time necessary for the maturation of an extraordinarily complex brain, rather than just for more time for learning (Fishbein, 1976). Comparisons clearly show the special features of human learning and the biological bases for "ease of learning" (Hamburg, 1963). The fact that *humans elaborate every biological predisposition* does not mean that the predispositions are not there.

Conclusion

In conclusion, I think that some social anthropologists in the future are *more likely* to advance the understanding of human behaviors if they know some human biology. This should lead to some social scientists becoming interested in biobehavioral problems, so that a new class of social scientists may develop the areas that link biology with social facts. But, if the development of the biosocial problems is to be useful and is not to repeat the mistakes of the past, it must start with the present and with a full comprehension of the complexity of human behavior. Our species elaborates everything, and only a very small part of this elaboration exists in the behaviors of any other animal. This is why comparisons must start with human beings, or the comparisons will inevitably be reductionist and simplistic. This is why linguistic behavior must be separated from the communication systems of other animals, because linguistic communication and cognition are the basis for the elaboration.

Finally, human evolution is a fascinating subject, and I think that it should be a part of everyone's general education. But evolutionary theories are complex, changing, and far broader than the part stressed in sociobiology. There is no theory that will predict behavior, but theories will help guide research and may help us to understand the results of the evolutionary process.

4 *Gerald Holton*

The New Synthesis?

*T*he invitation to share some thoughts on sociobiology has turned out to be a temptation too difficult to resist, despite all reservations. While I must base my remarks largely on a reading of the accessible literature—primarily E. O. Wilson's writings and commentaries on them—the obvious dangers for me and for anyone who is not an active researcher in biology are somewhat decreased because I have had an opportunity to check my preliminary conclusions with a diverse group of scholars in biological fields.

I shall be addressing three related questions: "What are the aims and claims of contemporary sociobiology?" "How does the enterprise fit into the history of ideas?" and "Does sociobiology have the earmarks of being indeed the beginning of a major synthesis?"

First I must make a distinction. There are really two pursuits, both referred to by the same term, *sociobiology* (defined by Wilson, 1975a, p. 595, as "the systematic study of the biological basis of all social behavior") and often indiscriminately merged in all discussions. In discussing the two pursuits, I will use the term "discipline," since Wilson and others have pointed out that it is too early to apply the word "theory." One of the two pursuits is what I would call the *Special* (or *Restricted*) Discipline; the other is the *Gen-*

eral Discipline. The former deals with animals below human beings. Wilson's book *Sociobiology: The New Synthesis* (1975a) devotes to the Special Discipline about 90 percent of the text pages and all but a handful of the approximately 2,500 references to research papers. And there seems to be little doubt that, in the sense of the Special Discipline, sociobiology "works" for large areas of animals exhibiting social behavior—perhaps best for social insects but, to a greater or lesser degree, also for others, from slime molds and corals to nonhuman primates. Thus, many specific observable and measurable aspects of behavior are correlated with genetic factors. To be sure, as in any growing scientific field, there should be and are vigorous debates about detailed observations and conclusions; for example, to what extent is the relative investment in the care of offspring influenced by the degree of genetic relatedness of individuals in the order Hymenoptera (wasps, ants, and bees)? Compare, for example, Trivers and Hare (1976) and Alexander and Sherman (1977). Even the assumptions underlying the concept of "parental investment" are being reexamined (J. Yellin and P.A. Samuelson, private communication). But the Special Discipline promises to mature soon into a Special *Theory* that may explain much of the observable social behavior of animals below humans. This by itself is no mean promise, not least because of the large number (an estimated 10,000) and the staggering biological diversity of social species that exist on this planet. One is led to expect to attain the use of one coherent corpus of variables and one theory to predict aspects of the social behavior of nonhuman animals from a knowledge of population parameters (demographic information concerning population growth and age structure) combined with information on the behavioral constraints imposed by the genetic constitution of the species. Such an achievement would surely be counted among the major advances of science, even if not a single word of it applied to the case of human beings. (One may add that if such a discontinuity in the application to humans were discovered in principle, that discovery in turn would constitute a major mystery for science.)

While the Special Discipline attracts by far the largest investment of energy of researchers in the field of sociobiology, the major focus of attention from those outside the field is the General

Discipline, which extends the promise and the program one crucial step further—to humans. It is just this inclusion of *human* sociobiology and the application to humans of otherwise well-established components of evolutionary theory that has transformed into a challenge what otherwise might have continued to be regarded as a specialty with limited interest. The challenge is signaled immediately in the subtitle of Wilson's book (*The New Synthesis*) and in explicit statements such as the following: "For the present, [sociobiology] focuses on animal societies, their population structure, castes, and communication, together with all of the physiology underlying the social adaptations. But the discipline is also concerned with the social behavior of early man and the adaptive features of organization in the most primitive contemporary human societies" (Wilson, 1975a, p. 4).

The extension to *modern* human beings, thought previously to be characterized in a qualitatively different way (for example, by an "open program"), is immediately indicated to be only a matter of time: "It may not be too much to say that sociology and the other social sciences, as well as the humanities, are the last branches of biology waiting to be included in the Modern Synthesis [neo-Darwinist evolutionary theory]. One of the functions of sociobiology, then, is to reformulate the foundations of the social sciences in a way that draws these subjects into the Modern Synthesis" (Wilson, 1975a, p. 4).

Wilson does not claim that all this is already happening. Although it may be plausible to *expect* continuity across neighboring species on the basis of particular findings in disciplines such as physiology, psychology, genetics, and demography, the general program is only sketched, and its supporting data for human sociobiology are few (in a variety of contexts—for example, incest taboo, infanticide, hypergamy, mental retardation and schizophrenia, and the biochemical basis of some behavioral mutations). But the driving force comes not only from such data; it comes also from an old dream: "The dream has been to bring biology—as a science, not simply as a source of unconnected facts—into conjunction with psychology, anthropology, and sociology and to make it part of the foundation of the social sciences. That goal may now be at least feasible, it not actually in sight. . . . It is hoped that knowledge of the subject will assist in identifying the origin and meaning of human values, from which all ethical pronouncements and much of political practice flow" (Wilson, 1975a, pp. 11, 12).

From time to time, Wilson is careful to ask for patience. Near the beginning of his book (p. 5), he points out that "the formulation of a theory of sociobiology constitutes, in my opinion, one of the great manageable problems of biology for the next twenty or thirty years," and he ends the book (p. 575) with the estimate that it may take as long as a hundred years. These cautionary words are all too easily lost sight of because of the emphasis given to the basic program of identifying and using the postulate of continuities across species—for example, that "the individual organism is only [the genes'] vehicle, part of an elaborate device to preserve and spread them with the least possible biochemical perturbation" (p. 3); that "in order to explain ethics and ethical philosophers" one must understand the role of natural selection in evolution, such as the connection between kinship and altruistic behavior; that "the hypothalamic-limbic complex of a highly social species, such as man, 'knows' or, more precisely . . . has been programmed to perform as if it knows; that its underlying genes will be proliferated maximally only if it orchestrates behavioral responses that bring into play an efficient mixture of personal survival, reproduction, and altruism" (p. 4); or, again, that "In this macroscopic view, the humanities and social sciences shrink to specialized branches of biology; history, biography, and fiction are the research protocols of human ethology; and anthropology and sociology together constitute the sociobiology of a single primate species" (p. 547).

To name only fields to which Wilson himself refers—the General Discipline is to lead to a synthesis, across all social species from colonial bacteria to human beings, of evolutionary biology, genetics, biochemistry, and ethology, and, specifically for humans, also of anthropology, psychology, sociology, the humanities, and ethics. One is led to expect a mutual accommodation of such conceptions as bonding, sex, division of labor, communication, territoriality, patriotism, warfare, learning, aggression, fear, altruism, and the structure of DNA. Indeed, what has been left out of this projected synthesis makes a very short list—chiefly the notions of the transcendental and of (undetermined) free will. But what has been left out will turn out to be of more than passing significance.

It may not be inappropriate to inject here a personal opinion. Regardless of the success this program may ultimately have, I find it admirable for four reasons. First, science needs more such wide-ranging, "risky" intellectual efforts to balance our usual fare of small additions to the sandheap of individual analytical results. Sociobiology is attempting to become a "theory of principle" (a theory covering a wide domain in which a large variety of verifiable results are obtained deductively from a few secured postulates), rather than a phenomenological theory (the more common type of theory, characterized by narrow domain and by many ad hoc explanations via plausible inductions, with short chains between the observations and the conceptual material). Second, even if it fails eventually (as all systems do), the challenge that sociobiology is throwing down before the neighboring disciplines can have a strong, perhaps transforming, effect on some of them—although not necessarily along the lines envisaged by sociobiologists. And that is one way in which progress is traditionally made. Third, Wilson's *Sociobiology* and related writings by proponents may be viewed as significant cultural artifacts in their own right, because they represent a world view characterizing this part of the twentieth century—for example, in their plea for a sophisticated form of flexible, almost stochastic, predeterminism and materialism; in their apparently dispassionate concern with a secularized ethic; in their accent on rationality and their underemphasis on affect and symbolic forms. In short, with all their limitations, they exemplify what is widely considered to be some of the best thinking today. Fourth and last, but not least, the discussion of sociobiology among biologists and other scholars can and should present opportunities for the difficult and all-too-often neglected task of exploring the possible impacts scientific work may have on ethics and human values.

Outraged Sensibilities

The opportunity for such assessments has only begun to be taken seriously. So far, the scene has been dominated by expressions of outraged sensibility—often but by no means only on the part of nonbiologists—triggered, although not fully explainable, by the type of statements on sociobiology made by its protagonists.

These responses themselves are well worth studying as events in the history of science. This is not the place for such a study, but it would be useful to identify, at least in a general way, a few reasons for the sense of discomfort produced in some quarters by the very discussion of sociobiology. This reaction has been so strong that it has occasionally verged on becoming a case of "limitation of scientific inquiry." For, as Lewin (1976, pp. 344–345) put it, "There is no doubt that many people have steered clear of the issue for fear of being labeled either as neo-Nazi or a hysterical radical."[1]

Opposition to sociobiology takes two forms. One is the understandable controversy within the specialty field that must test the claims of new proposals. A more visible opposition, ranging in intensity from polite disapproval to organized disruptions of meetings, focuses chiefly on the General rather than the Special Discipline; usually, the opponents do not claim to disprove data or conclusions as in the normal process of theory validation—no doubt in part because of the early state of sociobiology. At bottom, the more vehement objections seem to have one or more of the following three separable but not independent bases.

First, it would be wrong to deny that scientists and scholars,

---

[1] A petition condemning research in sociobiology was presented by a group to the council of the American Sociological Association at two of its meetings in September 1977. At one of these, the council, after study, passed the following resolution:

> The council expresses concern about the problems of overgeneralization from preliminary and tentative findings in sociobiology and about unwarranted ethical and moral conclusions that such overgeneralizations may have been used to buttress. The sociological community is therefore urged to examine the work of sociobiologists with due care and to publish appropriate comments and critiques of these studies with the same penetrating and critical perspectives as are applied to other areas of investigation. However, council takes the position that (1) it is inappropriate for council to decide by majority vote intellectual issues that are more properly considered by the informed community on the basis of scientific evidence; (2) it is inappropriate for council to condemn a varied and multifaceted area of investigation on the basis of the alleged shortcomings of some of its proponents; and (3) it is inappropriate for council to declare a field of inquiry illegitimate on the basis of the anticipated consequences of the application of its principles.

There is, of course, also a large and growing literature of informative critiques and assessments; the most recent works at hand are papers by Alexander, Lewontin, Kauffman, and Ruse, in Suppe and Asquith (Eds.), *PSA 1976*. Vol. 2: *Symposia* (East Lansing, Michigan: Philosophy of Science Association, 1977), and a long book review of Wilson's book by Dupree in *Minerva* (1977).

like other mortals, can be influenced by their "gut reactions." A good part of the reactions I myself have heard show this component. If the program of sociobiology at its most ambitious were to work, it is argued, it threatens "to short-circuit the person in the egg-egg cycle"; once more the progress of science would "objectivize the subjective" and "rationalize at the cost of affect and passion." We would then have a "clockwork model of man" as the "triumph of reductionistic scientism," bisecting human nature. In a replay of the seventeenth-century separation of primary and secondary qualities, we would be "casting away the qualitative, the ambiguous, the complex, and the artistic"—in short, much of what "makes each person unique with respect to any other person and that which makes mankind unique with respect to other species." When *Time* magazine (1977, pp. 54–63) recently published a long and, on the whole, rather balanced cover story on sociobiology, entitled "Why You Do What You Do—Sociobiology: A New Theory of Behavior" by Ruth Mehrtens Galvin, it chose for a cover, as a kind of emotional shorthand, a picture of two puppets representing a young man and a young woman, looking helplessly and vacantly past each other as they dangled on their strings in the frozen gesture of an abortive embrace.

Nor is Wilson insensitive to the dangers. In the final section of his book, he speaks of the purposes to which evolutionary sociobiology might be put in the future:

> If the decision is taken to mold cultures to fit the requirements of the ecological steady state, some behaviors can be altered experientially without emotional damage or loss in creativity. Others cannot. Uncertainty in this matter means that Skinner's dream of a culture predesigned for happiness will surely have to wait for the new neurobiology. A genetically accurate and hence completely fair code of ethics must also wait.
>
> The second contribution of evolutionary sociobiology will be to monitor the genetic basis of social behavior. . . . If the planned society—the creation of which seems inevitable in the coming century—were to deliberately steer its members past those stresses and conflicts that once gave the destructive phenotypes their Darwinian edge, the other phenotypes might dwindle with them. In this, the ultimate genetic sense, social control would rob man of his humanity. [Wilson, 1975a, p. 575][2]

[2]Wilson adds, at the very end, "When we have progressed enough to explain ourselves in these mechanistic terms and the social sciences

Second, expressions of this sort, by the critics and by Wilson himself, reveal a rather widespread *fear of abuse*. By themselves, fears do not form a rational basis for deciding where sociobiology will be heading, not to speak of whether the inquiry should be limited even if it could be. But even if not rational, they can be reasonable extrapolations of ominous present trends. One of humanity's oldest preoccupations is the pursuit of vice and folly; in our time, advances of science and technology have been eagerly incor-

---

come to full flower, the result might be hard to accept. It seems appropriate therefore to close this book as it began, with the foreboding insight of Albert Camus: 'A world that can be explained even with bad reasons is a familiar world. But, on the other hand, in a universe divested of illusions and lights, man feels an alien, a stranger. His exile is without remedy, since he is deprived of the memory of a lost home or the hope of a promised land.' This, unfortunately, is true. But we still have another hundred years" (p. 575).

Of those who responded to a draft version of the paper I had circulated, one of the most interesting letters came from Alexander Morin of the National Science Foundation. Saying (in part) that there may be more at work in the "gut reaction" against sociobiology than suggested, he went on:

Why does it arouse such passionate opposition, even among people who in other fields of inquiry are (or appear to be) dispassionate in their scientific consideration of science? Because what we are seeing, I think, is *not* a scientific response to evidence but a doctrinal response to heresy. Specifically, to three kinds of heresy:

First, sociobiology denies two essential elements of the Greco-Judaic tradition: mind-body dualism and the special creation of man. These notions persist with great strength. Even among those who presumably regard themselves as atheists liberated from ancient dogma, there is a persistent effort continuously to redefine those forms of behavior, independent of biological forces, that distinguish man from all other forms of life. Hence, sociobiology is sinful.

Second, sociobiology violates Durkheim's injunction—which is bedrock in the training of social scientists in this country—that social phenomena can only be explained in terms of social variables. I think originally this was an ingenious nineteenth-century bourgeois defense against the Marxist and Spencerian monisms. But the exclusion from social theory of independent variables exogenous to social systems (biology, ownership of the means of production, environment, whatever) is now needed to protect the social scientists from intrusion by outsiders, regardless of their ideological persuasion. Hence, sociobiology is "wrong."

Third, sociobiology has implications (and may ultimately provide conclusions) that contradict the notion of the perfectibility of man, which is a fundamental assumption in the dominant political ideology of science as it has developed in the West. It is assumed that with the help of science, we can create a social order such that all men will (at least for all practical purposes) be equal. Insofar as sociobiology suggests limits to this process, forces that cannot be overcome by tinkering with the social order, it is dangerous to established belief systems. [Quoted by permission] porated into that project. More evidence piles up day after day—the insanity of heaping higher the mountain of ever more fiendish weapon systems, the behavior control "experiments" of secret police on all sides of the oceans, the callous discharge into the environment of harmful by-products of industrial processes, and so forth. If greed and sadism have managed to benefit from the labors of scientific workers, it is reasonable to fear that other widely diffused human tendencies, such as xenophobia, racism, and the like, could fashion themselves some protective "scientific" cloak. One remembers the abuse of Social Darwinism in such fields as economics, immigration policy, and eugenics in Edwardian Britain and elsewhere—not to conjure up its deadly pervasion by the Nazis with the full cooperation of German doctors, scientists, lawyers, and administrators. Nor have the less obscene forms of Social Darwinism been completely conquered yet. The sheer instinct of self-preservation may be sufficient to account for the fact that people are suspicious as never before about any new scientific theory or technological development that might enlarge the potential for the control of human behavior. We ask, "Control by whom? According to whose values? For whose benefit and at whose risk? With what institutional constraint?"

The loudest protests I have heard leveled against sociobiology do not claim that any of the feared abuses have already occurred. So far, no specific proposal for basing social policies on current sociobiological knowledge has surfaced. The fears tend to refer only by analogy to what may have happened in related fields. To be sure, it is a new and difficult calculus: Some modern victims of the perversion of science and technology are all too easily identified; others are not. (How would one prove to have been personally harmed by an escalation in the balance of terror? Or by feeling more and more like those puppets on the strings?) In this circumstance, sentiment can make itself felt—and over the past few years sentiment has been shifting, as it did in the handling of food additives. The Food and Drug Administration used to label most of them simply as GRAS ("generally regarded as safe"), but now additives are considered guilty until having been proved innocent. Similarly, in many quarters those aspects of science and technology that have health- or behavior-affecting potential are no longer

GRAS, from recombinant DNA research to nuclear engineering. The new phrase *limits of scientific inquiry* characterizes the whole movement. The skepticism about near dangers is no longer tranquilized by the promise of more distant rewards—even by the promise of exposing, in the long run, those earlier biological and social adaptations that once may have been functional but that now are disastrous for mankind. Wilson (1975a, pp. 24–25) himself writes: "For example, the tendency to expand at the expense of territorial neighbors might well be in human genes, having been advantageous to our ancestors through evolutionary time, but it would lead to global suicide now. To rear as many healthy children as possible has been the road to security in many cultures and periods of history, but with the world's population brimming over it is now the way to environmental disaster. To an increasing degree, we are forced to make moral decisions that directly influence the future of the human species. Soon we may have to pick and choose among the emotional guides that we have inherited, and determine those that should be followed and those that should be sublimated or redirected so that our behavioral patterns will both conform with biological principles and foster the growth of the human spirit." It would have been improbable that any form of sociobiology applicable to human beings, coming on the scene in this part of the century, would have been exempt from consequences of the current strain of pessimism.

Third, yet another "reasonable" and expectable type of adverse response to sociobiology may be identified as *territoriality and dichotomization*. A number of intellectuals in fields neighboring on sociobiology are concerned about what they perceive to be grand imperialistic designs on their area, and they are not calmed by the casual disclaimer that the success of the program may be a hundred years off. Comments I have heard made by academics under this heading have contained such accusations as that sociobiology "trivializes" the work of social scientists by "disaggregation" and shifts the "battlefield" to an entirely inappropriate area, that the ontogeny of human behavior must continue to be based first of all in the analysis of childhood experiences, and that the whole enterprise is implausible because one cannot imagine the chain of "intermediate causal steps" necessary for understanding how heritability expresses itself operationally in behavior.

Wilson himself cannot have been entirely unprepared for the professional resistance. When he announced (1975a, p. 6) that such fields as ethology and comparative psychology "are destined to be cannibalized" because the future cannot rest with their "ad hoc terminology, crude models, and curve fitting," he added "I hope not too many scholars in ethology and psychology will be offended by this vision." In his more recent writings, Wilson has done little to calm colleagues in neighboring territories. On the contrary, his essay "Biology and the Social Sciences" (1977) contains a direct attack on the separation between fields having adjacent levels of organization. Going far beyond the so-called Modern Synthesis of Mendelian genetics and biochemistry, he envisages a "juncture" of neurobiology and sociobiology with social science. He focuses on the creative tension between neighboring fields whose relationship makes them act as "antidisciplines": "By today's standards, a broad scholar can be described as one who is a student of three subjects: his discipline, the lower antidiscipline, and the subject to which his specialty stands as antidiscipline [at the next level of organization]. A well-rounded neurophysiologist, for example, is deeply involved in the microstructure and behavior of single cells, but he also understands the molecular basis of electrical and chemical transmission, and he hopes to explain enough of neuron systems to help account for the more elementary patterns of animal behavior" (Wilson, 1977, p. 128).

In the evolution of molecular biology, "progress over a large part of biology was fueled by a competition among the various attitudes and themata derived from biology and chemistry—the discipline and its antidiscipline" (Wilson, 1977, p. 129). Wilson feels that a similar process will eventually occur for sociobiology as the biologist glimpses the "reverse side of the social sciences." For example, economics will be understood from so general a perspective that the conventional treatment of the subject becomes merely "the description of economic behavior in one mammalian species with a limited range of the biological state variables" (1977, p. 36), rather than the actions of human beings in the marketplace.

It is not surprising to find assertions of territorial claims in the replies to Wilson by members of other disciplines. But neither these assertions nor suspicions about the validity of the new methodology with its high ambition reveal the passions involved. Be-

tween sociobiologists and their opponents, there is also evidence
of a clash of fundamentally differing world views. Most intellectuals find it difficult to hold and juxtapose in their minds two still
developing systems with overlapping jurisdictions, the more so if
the systems are based on incommensurate assumptions. This difficulty produces the cultural equivalent of a cognitive clash between sociobiology and the other approaches to understanding
human behavior (as in humanistic psychology, where ambiguity,
complexity, and confusion are handled quite differently). In this
clash, the solution is all too often found in dichotomization, in the
tendency to exclude all but one system instead of attempting to
hold two or more systems in parallel.

Some Precursors

Just as the opponents of sociobiology can cite plausible motivations for their pessimism, the proponents have their own case
for optimism. To deepen our understanding of the aims and
claims, the powers and limits, of contemporary sociobiology, we
must now ask how the enterprise fits into the history of ideas. The
whole field of research and the motivating spirit behind sociobiology are not the products of the last decade or two, as a citation
analysis might lead one to believe. On the contrary, it is part of a
long development. Sociobiology too has its phylogeny and was already well established in the middle of the nineteenth century, at
the time when the mechanists and vitalists were doing battle.[3] In
1845, a group of young physiologists, among them Helmholtz and
Dubois-Reymond, swore an oath to account for all bodily processes
in physical-chemical terms. They did not prohibit all metaphysical discussions of that science, but merely declared, in Dubois-
Reymond's famous phrase, *ignorabimus*—that is, we shall never
know the great world riddles, except those portions that reveal
themselves within mechanistic science.

This group was distinguishable from a parallel, but more

---

[3]Of the many accounts, I refer to the excellent, brief one by Donald
Fleming in his introduction to the reissue of Jacques Loeb's *The Mechanistic
Conception of Life* ([1912] 1964). Also see Everett Mendelsohn (1974).

extreme, group of experimental biologists and medical materialists who may be called the "nothing-but" school. To them, all things were to be reduced to a homogeneous mechanistic scheme, including the world riddles despaired of by the others. This naturally led them to attack the established order, the alliance between church and state, and all the other impedimenta to radical progress within science and without. Not surprisingly, many of them were socialists and visionary fighters for social justice. For example, Rudolf Virchow, one of the sympathizers, supported the German Revolution of 1848 and became the chief of the liberal opposition to Bismarck. (See also Gregory, 1977.) From the present perspective, the medical materialists and the Helmholtz group were far closer to each other than to any of their common enemies; they were, for example, united in being antitranscendentalists.

To me, the most interesting figure among all these was the biologist Ernst Haeckel. Haeckel was one of the major influences in bringing Darwinism to Germany. Long after his death, this Darwinism was twisted in an effort to lend respectability to programs of euthanasia and genocide. A fiery materialist and socialist, Haeckel scoffed at all myth mongers and offered a complete world view based on evolution and monism (unity of mind and matter) that would solve all puzzles. The turbulent book he wrote in 1899—toward the end of his career but at the height of his fame—was in fact titled simply *The Riddle of the Universe* (*Die Welträtsel*). It swept over Europe like a crusade against mystification, against what he regarded as "the untruth foisted on the people by their spiritual and economic masters." Science was to triumph over theology by spreading the gospel of evolution infused with a modicum of panpsychism. Haeckel's chief point was that there was a unity of the inorganic and the organic world, grounded in the laws of conservation of matter and energy (what he called "the law of substance").

It was indeed a replay, complete in many details, of an ancient message. Here it is first in the words of Lucretius, introducing the world view of the earliest Greek atomists. "I will essay to discourse to you of the most high system of heaven and the gods, and will open up the first beginnings of things, out of which nature gives birth to all things and increase and nourishment. . . . Nothing

is ever gotten out of nothing by divine power. Fear in sooth takes
such a hold of all mortals because they see many operations go on
in earth and heaven, the causes of which they can in no way understand, believing them therefore to be done by divine power.
For these reasons, when we shall have seen that nothing can be
produced from nothing, we shall then more correctly ascertain
that which we are pursuing, both the elements out of which everything can be produced and the manner in which all things are done
without the hands of the gods" (Lucretius, [c. 99–55 B.C.] 1969,
pp. 2–5).

The promise of external persistence and of a guiding polestar was vivid in the sweeping and reassuring chapter titles in

In Haeckel's own battle against such notions as personal immortality, the conventional belief in a creating God, or in the belief
in a mind or a purpose behind evolution, he did not have to refer
explicitly to Lucretius. Haeckel's sentences had their own grand,
Teutonic sweep: "All the particular advances of physics and chemistry yield in theoretical importance to the discovery of the great
law which brings them to one common focus, the law of substance.
This fundamental cosmic law establishes the eternal persistence of
matter and force, the unvarying constancy throughout the entire
universe. It has become the polestar that guides our monistic philosophy through the mighty labyrinth to a solution of the world
problem" (Haeckel, [1899] 1929, p. 3).

The promise of external persistence and of a guiding polestar was vivid in the sweeping and reassuring chapter titles in
Haeckel's book: "The History of Our Species," "The Phylogeny of
the Soul," "Consciousness," "Immortality," "The Evolution of the
World," "The Unity of Nature," "Our Monistic Ethics," and,
finally, "The Solution of the World Problems." In comparison,
Wilson's book is an exercise in understatement and scientific objectivity. I doubt that it is able to arouse a small fraction of the
hopes and fears that Haeckel's book did for about half a century.

Another precursor of Wilson is Jacques Loeb, the author of
The Mechanistic Conception of Life ([1912] 1969). Born in 1859, he
was a scientist in the old style of philosopher and social innovator,
certain that scientific findings might lead directly to political and
social development. Influenced by Schopenhauer (as were so many
others of his generation), he seems to have turned to biology in

order to find evidence against the conception of the freedom of the will. Perhaps his best work was on animal tropism, the involuntary movements imposed by such environmental conditions as light on organisms; he considered it a model for understanding behavior in terms that avoid using the noxious conception of "will." The accomplishment for which he is most famous—artificial parthenogenesis by physical-chemical means—fell in the same category of scientific research findings with antitranscendental and antimetaphysical implications.

From 1911 on, cheered by the proof of the existence of molecules by Jean Perrin and others as the triumph of mechanistic philosophy, Loeb spoke and wrote on "the mechanistic conception of life" and published his book of that title in 1912. In it, as Donald Fleming observed in the introduction to the 1969 edition, Loeb reduced life to a physical-chemical phenomenon, free will to an illusion generated by tropistic causes, and religious faith to an absurdity. He proclaimed the total validity of mechanistic principles and derived from them a system of human ethics based on instincts whose unobstructed expression would rejuvenate world society. In his book, Loeb asked whether human "inner life"—the "wishes and hopes, efforts and struggles"—should be "amenable to a physical-chemical analysis" (Loeb, [1912] 1969, p. 28). And he answered yes, even if the proof would have to come from much research that still waited to be done: "For some of these instincts, the chemical basis is at least sufficiently indicated to arouse the hope that the analysis, from the mechanistic point of view, is only a question of time" ([1912] 1969, p. 32).

In the last pages of this work, just as in Haeckel's and in Wilson's, Loeb has a section entitled "Ethics." Here is a passage: "We eat, drink, and reproduce, not because mankind has reached an agreement that this is desirable, but because, machinelike, we are compelled to do so. We are active because we are compelled to be so by processes in our central nervous system. . . . The mother loves and cares for her children, not because metaphysicians had the idea that this was desirable, but because the instinct of taking care of the young is inherited just as distinctly as the morphological characters of the female body. . . . Not only is the mechanistic conception of life compatible with ethics: It seems the only conception of life which can lead to an understanding of the source of ethics" (Loeb, [1912] 1969, p. 33).

In comparison, Wilson's is a more sober, more scientifically grounded effort. Ironically, and partly for this reason, it will not have the same popularity that these predecessors had.

Evaluating the Potential for Synthesis

Against this background, we can now evaluate the inherent claim of sociobiology that it produces the "New Synthesis" (or *any* synthesis). What, indeed, is the *structure* of a synthesis, and how does sociobiology correspond to it?

I view synthesis and analysis as methodological themata, synthesis being one component of the thematic pair, the other being analysis. (I have discussed these conceptions at greater length in *Scientific Imagination: Case Studies,* Holton, 1978.) The term *synthesis,* of course, brings to mind certain methodological practices in the works of philosophers since Plato. But it is necessary to distinguish between four general meanings of the term: (1) in the *reconstitutional* sense (for example, where an analysis followed by a synthesis reestablishes the original condition), (2) in the *transformational* sense (for example, where the application of analysis and synthesis advances one to a qualitatively new level, whether in a given specialty field, such as biology, or in two specialty fields, such as biology and sociology); (3) in the *judgmental* sense (as in the Kantian categories and their modern critiques), and (4) in the general, *cultural* sense.

To specify the properties of a synthesis in operational terms, we select a body of work that is, beyond challenge, a historic synthesis, and use it both to identify the structure of a working synthesis in science and to measure how close sociobiology may be to the model. The Newtonian synthesis (the historic unification of celestial and terrestrial physics) is probably the most distinguished example; while referring to it to compare the half dozen major structural elements of any synthesis, we shall keep in mind that we thereby calibrate, so to speak, the top reading on that kind of thermometer.

Historic Roots. Almost by definition, a synthesis has roots in the history of the fields within which it produces coherence. For the Newtonian synthesis, one of these roots reaches back to the grand scheme of Thales of Miletus, the other to Pythagoras of Crotona. The former was essentially positivistic and materialistic, with a certain resemblance to modern empiricism, while the latter was metaphysical and formalistic. It is significant that these systems, which came into Western culture at about the same time—2,500 years ago—were impelled by the persisting drive to find basic unity underlying the diversity of all experience but nevertheless were diametrically opposite in assumption and appeared mutually exclusive in content.

From each of these two schools, a separate chain of distinguished followers emerged over the next centuries. Aristotle stands at a pivotal position in the history of thought, in part because he was the first major thinker who was not a follower of only one of the two main trends and because he made a powerful attempt to adapt elements from both of the antithetical systems in a new synthesis. Nothing even faintly analogous was done successfully in natural philosophy until the joining by Kepler and Galileo of neo-Platonic and materialistic conceptions. Newton's synthesis, then, was the last grand bridging of the materialistic-positivistic tradition and the formalistic-metaphysical tradition in natural philosophy. Later attempts were restricted to narrowly delimited fields within the physical sciences. Thus Faraday's central theme, in his research on relations between gravity and electricity, was what he called "the long-standing persuasion that all the forces of nature are mutually dependent, having one origin, or rather being different manifestations of one fundamental power." To this day, this is the Holy Grail of theoretical physicists, who try to find *one* force to explain the gravitational and the electromagnetic, the weak and the strong interactions.

Turning to sociobiology, we find that it too has a distinguished phylogeny: It is in fact the current terminal point on a trajectory or proliferation of system builders issuing primarily from the materialistic-mechanistic and antimetaphysical school of Thales of Miletus and his followers—Anaximander, Anaximenes, Heraclitus, Leucippus, Democritus, Anaxagoras, and Lucretius.

The more recent successors developed from the trajectory of these physiologues, teaching "disenchanted" or "positive" explanations of nature's phenomena, are (in varying degrees) Newton, Vico, Laplace, D'Alembert, Condorcet, Comté, Darwin, Helmholtz, Dubois-Reymond, Spencer, T. H. Huxley, Haeckel, Loeb, Mach, Julian Huxley, Haldane, the early Lysenko, and Schrödinger (in his book *What Is Life?*, 1967).

If we look beyond their many differences, they all share fundamental ambitions, approaches, and themes. For example, the matrix of social values and the moral base are taken not as *a priori* but as susceptible of explanation within a materialistic world view. These natural philosophers tend to opt for continuity instead of uniqueness, for unity rather than discreteness. In modern sociobiology, the old theme of classical physical causality persists, although modified and recast in terms of tendencies and "potentials"—an even older, yet still current, theme. Many of the moderns are social innovators and opt for an essentially optimistic and liberal political stance.

Thus, looking at this aspect of syntheses in general—synthesis as the climactic achievement of a long-term trajectory—the ambition of sociobiology is entirely recognizable.

Inclusion and Exclusion of Elements. The raw materials from which a synthesis must be fashioned are individual, seemingly disparate elements or separate classes of entities. Thus the Newtonian laws govern the motions of objects from atomic to galactic size. Yet Newton also specifically excluded large sets of elements from his synthesis—not only "occult qualities" that were no longer desired but also light and its propagation, chemical reactions, much of fluid mechanics and the theory of elasticity, sensations in the human body, and the properties of the ether.

Sociobiology seems to be in danger of not knowing how to exclude explicitly some tempting candidate elements. One has the impression that the range of behaviors, traits, and details clamoring for inclusion is enormously large. To be sure, history reminds us that exclusion very frequently is not, and perhaps cannot be, an *a priori* conscious decision but can only come at the end of a long series of unsuccessful attempts at inclusion. That is, exclusions are the result of the discovery of "impotency principles." And to find those one needs time.

A First Principle. After Newton, nothing so basic as the intuition of a universal law of gravitation furnishing the first principle on which to build a system will perhaps ever be granted to another synthesist. But sociobiology does make several basic, fundamental postulates—for example, the central theorem that animals behave so as to optimize their inclusive fitness; that there is some molecular basis of behavior (that the genes "program the potentials"); that for all phenotypes, including behavior, there is selection by interaction of genes and environment; and that there is a continuity of mammalian traits in humans. For the theory's eventual success in the large sense, it would seem necessary to postulate explicitly the smallest number of independent statements and, insofar as possible, to exhibit the role of parsimony and necessity among those postulates that do remain. (I am of course aware that some biologists may well object to this criterion, trained as they are to be more tolerant and respectful of complexity than are physicists.)

We also know from the study of earlier scientific advances that the formulation of powerful "first principles" often had to wait for the formulation of new concepts (energy, valance, invariance, quantization, complementarity). New terms, new metaphors, a new language parallel new generalizations. They cannot, of course, be identified in advance of the pressure for fruitful hypothesis; nevertheless, one might speculate on the type of additional concepts that a general theory of sociobiology may require.

An example would be one single conception that can handle simultaneously the opposite notions of potentiation and determination—something like a flexibly constraining field that contains decision vertices having probabilistic parameters. Perhaps new terms are also needed for those distinctive and preferably quantifiable human traits (if any) that are *not* shared with other species (for example, similar to the distinction between roles in human societies versus castes in social insects). Terms such as *altruism, slavery,* and *warfare,* when applied to nonhuman species, may turn out to have exhausted their usefulness and might now be recast, to

avoid unnecessary limitations or at least the charge of undue anthropomorphic connotation.[4] Conversely, since the bridges leading
from the study of social insects to that of humans are still narrow
and weak, they should not appear to be able to carry larger loads
merely because of possible confusions of word meanings.

   Cohesion of a General System. Again, since Newton (with the
possible exception of General Relativity) no system with such general coherence, no such deductive testable schema based on one
or a few principles, can be expected to arise again in our time—
least of all in a theory that is still under construction before our
very eyes. It will be a long time before we see the equivalent of the
deduction (and hence "explanation") of Kepler's empirical laws.
Yet there are beginnings in sociobiology, such as efforts to understand the mating system, the size of families and colonies, and diffusion speeds. More such advances, and a good cataloguing of
them, will be needed to make this synthesis widely persuasive. The
current literature shows that this is the growing edge of the whole
effort.

   I recognize that workers in other sciences may be understandably depressed (or overly impressed) by the success under this
heading that the 300-year effort of modern physics has had at its
culmination, for that is the model that physics willy-nilly puts before the other sciences. The power of the deductive network produced in physics has been illustrated in a delightful article by Vic-
tor F. Weisskopf (1975). He starts by taking the magnitudes of six
physical constants known by measurement: the mass of the proton,
the mass and electric charge of the electron, the velocity of light,
Newton's gravitational constant, and the quantum of action of
Planck. He adds three or four fundamental laws (for example, de
Broglie's relations connecting particle momentum and particle energy with the wavelength and frequency and the Pauli exclusion
principle) and shows that one can then derive a host of different
and apparently quite unconnected facts that happen to be known

[4]On this point, see the interesting essay, largely supportive of sociobiology, by Donald T. Campbell, "On the Conflicts between Biological
and Social Evolution and between Psychology and Moral Tradition,"
American Psychologist 1975, *30* (12), 1103–1126.

to us separately by observation; for example, the size and energies of nuclei and the size of our sun and of similar stars. This is indeed fulfilling Newton's program, triumphantly. However, one must not superimpose the same expectation on the program of Lucretius and the related one of Wilson.

Demystification and Central "Image." It is well known that Newton had a profound philosophical impact on his contemporaries by his demonstration that quite "ordinary" and causal chains were at work in producing complex or frightening effects (tides and comets, respectively) and that the world of infinite change was explainable by the persistence of a very few simple laws that any schoolchild could memorize. By extending the reign of familiar terrestrial processes and showing them to operate throughout the knowable world, a single, almost hypnotic image could suggest itself—that of the universe as a majestic clockwork. Its visualizability at one mental "glance" was probably a powerful factor in its popular acceptance.

Sociobiology in its current version does hold out a similar promise of "explaining" complex or disturbing effects in the processes of human society, from homosexuality to warfare. Even if there should be only a quite partial delivery on that promise, the effects on the world view of our society would be enormous. Sociobiology does seem to lack, however, a central image, analogous to the vision of a clockwork conjured up by the Newtonian synthesis. There is not even a complex one such as Darwin's "tangled bank." The voluminous and painstaking work chronicled in Wilson's book and similar sources may never lend itself to such a feat of fruitful oversimplification.[5]

Prediction. Predictive capability, as in deducing correctly the return of a comet from Newton's laws, is usually regarded as the

---

[5] Another of my correspondents kindly suggested, at this point in the draft, "It seems to me sociobiology does have an 'image' and a quite provocative one and that this has a great deal to do with its acceptance or rejection by various people. It isn't a simple image (as you point out), but it is a dramatic and compelling one nonetheless. That image is the appeal, in dramatic terms, to the human dilemma of selfishness and altruism, to the rooting of that dilemma deep in human nature, and to the agony and quandary that it causes. This is as evocative a view of human nature as

ultimate test of how "scientific" a synthesis is—the "harder" the science, the prouder and more confident it is. In this respect, sociobiology seems to be in only an early stage of development. However, we may be touching here on the long debate concerning essential differences between the biological and the physical sciences rather than on a *sine qua non* of scientific synthesis as such. For, at the very least, a far greater degree of complexity is built into biological systems by virtue of the necessary connection of each function with the organism's history on the one hand and with its environment on the other.

Cultural Reach. The claim of the Newtonian synthesis as a powerful exemplar of a cultural synthesis that changed civilization has been amply documented. (If one were allowed only a single example, an analysis of the role of Newtonian philosophy in Thomas Jefferson's draft of the American Declaration of Independence might suffice.)

By this measure, of course, the strategy with respect to sociobiology is, once more, patience. It does appear that both the proponents and the more vociferous and politically oriented opponents of sociobiology are united in the expectation that the New Synthesis of which Wilson speaks will be one that changes our culture. If it does not, the synthesis will still be one in the "transformational" sense.

The Promise Fulfilled?

One may conclude that on many, perhaps most, counts sociobiology has a fair chance eventually to bear out its promise—provided that more of the crucial elements are supplied in the ongoing research, notably with respect to cohesion and prediction—and thereby to advance from the Special Discipline to the General Theory. It can only be guessed how widely the New Synthesis will be welcomed by the adjacent disciplines. Nor can we predict whether

---

any Freud devised (and, not incidentally, very similar to Freud's). Wilson's view is intellectually complex, and a proper investigation of sociobiology in any species would be inordinately complex, but that doesn't make the popular image of it culturally less powerful."

the ideas of sociobiology will be as severely abused as were, on occasion, Darwin's and Einstein's—although one positive outcome of the sobering experience of the recent past for all scientists should be their greater watchfulness and, where necessary, activism to expose abuses forcefully as soon as they are actually identifiable.

The issue of *scientific validity* will of course be decided in the laboratories and in field research. If experimental evidence in favor of human sociobiology does turn out to be voluminous, varied, and positive, the founders of the field will no doubt be installed in the Pantheon. But it is a curious question whether Haeckel and Loeb and the others who will be waiting for them there will in fact approve of this new version of the ancient quest. No doubt they will like to hear evidence that the social behavior of animals can generally be linked, across all species, to the mechanism of natural selection. They will also be pleased that biology, anthropology, and many neighboring fields will have been shaken up in a fruitful way. But we can imagine they will at least raise an eyebrow that in our time the offspring of Lucretius no longer found theological opponents to engage head on and addressed themselves instead to the modern equivalent of the ancient seat of moral force—that is, to the social sciences.

In answer, the new arrivals will have to plead that the road to the New Jerusalem once more proved to be more difficult than had at first appeared and that it may be quite enough if we end up wiser about the behavior of people and other animals. Even in making a synthesis, there is a large and useful middle ground between complete success and failure.

5

John L. Fuller

Genes, Brains, and Behavior

$\mathcal{T}$he letter inviting me to participate in this book suggested that I consider the possibility of explaining whether "characteristic human cognitive behavior can be explained by reference to primary gene action." My first reaction was to decline with thanks on the ground that no one, certainly not I, could explain that relationship beyond the simple statement that, "It takes a human genotype to develop a human mind." However, I did accept because I liked the idea of bringing scientists and humanists together in a common intellectual task. After all, both science and the humanities depend upon a collection of forty-six chromosomes that work together to produce a brain that can ask questions about itself. Herein lies our human uniqueness.

A recent conference of philosophers and neuroscientists on the topic of consciousness and the brain demonstrated that the old issues of monism, parallelism, and dualism are still alive and that intelligent persons are far from agreement (Globus, Maxwell, and Savodnik, 1976). My charge, to go beyond the brain to the genes that direct its development, is even more difficult. It is, however, an issue raised by Wilson (1975a) in *Sociobiology: The New Synthesis*, for to him our nature and culture are the outcomes of natural selection operating on gene frequencies in populations.

Wilson's *Sociobiology* has aroused interest among scholars of many disciplines that is comparable to that generated by Skinner's *Beyond Freedom and Dignity* (1971) and perhaps even by Darwin's *On the Origin of Species* ([1859] 1964). All three books treat the human being as an animal—a very special kind of animal, to be sure, but a creature with no more divinity than any other species with which it shares this planet. Wilson specifically considers ethics as a branch of evolutionary biology; our cherished human values are to be judged in terms of their effect on the survival of the genes of those individuals who live according to our ethical precepts. Thus sociobiology is threatening both to those who view ethical principles as God-given eternal truths and to those social planners who have faith in their unlimited ability to reconstruct ethical and cultural systems through social conditioning. Wilson, while stressing the tremendous variability of human skills, interests, and forms of social organization, tells us that there is a human biogram that extends beyond the need to eat, drink, eliminate, and copulate. At the heart of the biogram are behavior patterns that are similar to, although less rigid than, the stereotyped displays of fish, birds, and other mammals. He implies that this human way of life is the product of genotypes that have been assembled through natural selection and that there are limits to its modifiability.

Genetics in Darwin, Wilson, and Skinner

Genetics plays an important role in both *On the Origin of Species* and *Sociobiology*. Darwin knew nothing of Mendelian or biochemical genetics; his theory of natural selection was based on the inheritance of form and function through the germ cells. Wilson is well acquainted with modern genetics, and he interweaves this knowledge with observations on the ecology and ethology of a great variety of living species. Skinner is generally considered to stress environmental explanations of behavior over genetic ones, but he devotes a chapter of *About Behaviorism* (1974) to innate behavior, the starting point for the conditioning process. And in his article "The Phylogeny and Ontogeny of Behavior," he argues that the evolutionary processes that shape the biogram of a species are analogous to the conditioning processes that shape its behavior (Skinner, 1966). Genes mutate into new forms; new genotypes (combinations of genes) are produced by the union of sperm and ovum. These new genes and new genotypes potentially affect all aspects of organisms, including their behavior. The effects may be neutral, negative, or positive with respect to enhancing reproductive fitness. Over generations, mutation, reassortment, and natural selection thus lead to the changes in the gene pool that we call *evolution*. Although natural selection is the great shaping force in this process, there must be preexisting genetic variation for it to be effective.

The principles of operant conditioning have a basic form similar to that of natural selection, although the two differ widely in mechanisms and time scale. An organism tends to increase the frequency of responses that prove to be rewarding. But before a response can be reinforced it must first be produced. Genes that facilitate the early emission of responses that are likely to be reinforced should be favored by selection. Thus natural selection and learning must affect each other on the phylogenetic time scale. Skinner sees no reason why some responses to the external environment should not come "ready-made," in the same sense that many responses to changes in the internal environment are considered to be "inherited" rather than acquired. Behaviorism at the individual level seems to be compatible with sociobiology at the species level. The big question for a human-oriented sociobiology is the degree of complexity that can be encoded in the genotype and the degree of flexibility that is built into the system.

The validity of the sociobiological model for human behavior requires demonstration of the following principles: (1) There must be extensive genetic variation in our species. (2) This genetic variation must have a demonstrable influence on behavior. (3) Gene-associated behavioral variation must have an effect on biological fitness—fitness being defined in terms of relative success in transmitting one's genes to one's descendants. The first of these principles requires little discussion. Human beings are so genetically diverse that it is unlikely that two genetically identical individuals have ever existed—except, of course, monozygotic cotwins. The second two principles are the concern of behavior genetics, a branch of science that studies many types of organisms, including *Homo sapiens.*

Behavior Genetics

The roots of behavior genetics antedate the modern scientific era. There are hints of a genetic theory of intelligence in Plato's *Republic,* where he argues for coeducational academies in which the most intelligent, comely, and healthy young men and women would be closely associated. Natural attraction between the sexes would suffice to ensure marriages among the best endowed, whose good traits would thus be transmitted to their children. From Plato through Galton's *Hereditary Genius* (1884) to the early part of the twentieth century, the inheritance of behavior was considered to be as obvious as the inheritance of physical traits. Doubts in this matter were expressed from time to time by individuals who recognized that human parents provide much more than genes to their offspring. The battle between hereditarians and environmentalists was bitter in the 1930s and has not completely disappeared in the 1970s, despite periodic attempts to declare that it is a pseudoproblem. From today's perspective, many of the arguments on both sides are naive and polemical rather than constructive. Two major errors in logic are the assumptions that a particular form of behavior is either learned or innate and that the demonstration of improvement with practice rules out genetic explanations for behavioral differences.

In the 1960s, a specialized area variously called *behavior genetics, behavioral genetics,* or *psychogenetics* became defined when biologists, psychologists, psychiatrists, and a few sociologists discovered that they had common interests in genes as a possible source of individual variation in behavioral traits. The science was built partly on quantitative genetics, which supplied Mendelian models for the inheritance of such continuously distributed traits as intelligence and personality and partly on the discovery of specific genetic bases for many types of mental retardation. Parallel with the work on humans, animal behavior was subjected to genetic analysis.

By this time, genes were no longer hypothetical constructs but were megamolecules of DNA that directed the synthesis of enzymes and other proteins. At the same time, advances in population genetics led to new concepts of human variability and new understanding of the evolutionary process. The development of an organism, both physical and behavioral, was now seen as a continuous coaction of genes and their products with an environment. So closely interlocked were these factors that it made no sense to say that one part of behavior was inherited and another part learned. One could, however, sometimes state with confidence that a portion of the phenotypic variance in a population was attributable to genotype and another portion to environment.

What has this new hybrid science to say about the interrelationships of genes, brains, and behavior? To start, I shall consider the relationship of genes to the brain, the organ of greatest interest to psychologists. How can we account for the construction of this extremely complex assembly of cells through the action of a batch of DNA and RNA molecules on a collection of amino acids? We can be sure that the genome does not contain a blueprint for the brain; there are not enough genes to direct each neuron to make proper connections with other neurons and to select an appropriate sensory or motor cell. The genomic instructions to a neuron cannot be in the form of a road map. Instead, they must be of this type: "Start growing now and follow the path of least resistance, and when you find an appropriate cell connect with it. How will you know when you have found the right cell? Its surface will feel right to you." Good summaries of developmental neurobiology can be found in several recent books (Jacobsen, 1970; Gottlieb, 1973a, 1973b, 1976). I confess to a feeling of amazement that the system works so well and that the brains of individual members of a species are so much alike. Actually, the brains of all members of a species are not identical. Striking genetic differences have been found in the brain structure of "normal" laboratory mice, but the psychological significance of these differences is still undetermined. Sometimes the coordination of brain development breaks down when a mutant gene directs the synthesis of a wrong molecule. But even in instances of gross genetic error much of the species-specific brain structure and behavior remains. A child with Down's syndrome is a retarded human—not a home-reared ape.

Most of us, aside from neurosurgeons, neuroscientists, and physiological psychologists, are more directly concerned with the development of behavior than with the development of the brain. We are also interested in the extent to which individual differences in behavior are attributable to genetic variation and in the range of possible phenotypes with which any given genotype can be associated. Unless behavioral variation is heritable, the whole structure of sociobiology falls to the ground. Behavioral geneticists take the long leap between genes and behavior with a certain amount of trepidation and considerable caution, but they have abundant evidence for functional relationships between them. Their first problem is the definition of a behavioral phenotype or psychophene. Most geneticists work with somatophenes that are defined by their structure and that can be subdivided into chemophenes and morphenes. Chemophenes are molecules identifiable by chemical or immunological analysis. They are primary gene products or their immediate derivatives. Morphenes are structures of a higher order, such as fingerprint patterns or the shape of one's nose. Molecular biology tells us a great deal about how genes regulate the production of chemophenes but tells us little about how they interact to specify morphenes. Psychophenes are not structures but processes whose nature is inferred from observations on living organisms. From behavioral data, we infer the existence of a relatively stable characteristic or trait of an organism. A rat that defecates frequently in an unfamiliar open field is "emotional"; a child who answers most of the questions on a standard test is highly "intelligent." We know, of course, that behavior involves transactions between an organism and its environment and that the testing situation must be carefully controlled if we are to detect stable psychophenes. It is useful to divide behavior genetics into two major areas: one dealing with the extent to which differences among individuals in a population are ascribable to their genes, the other dealing with the developmental processes intervening between genes and behavior patterns.

Behavior-Genetic Analysis of Social Behavior

By comparing the degree of resemblance of related individuals who have genes in common with the resemblance of nonrelatives, we can estimate how much of the variance in a population is attributable to heredity. The ratio of genetic to total phenotypic variance is called *heritability*. Space does not permit an exhaustive review of this much misunderstood and criticized term. The concept of heritability is neither as valueless as its detractors claim nor as useful—for humans, at least—as it has been in the practice of animal and plant breeding (Layzer, 1974). We can state with confidence that in animals there is overwhelming evidence for the heritability of many forms of social behavior, such as courtship, mating, caretaking, and aggression. All of these functions play a major role in sociobiological theory. Behavior-genetic analysis in animals is facilitated by our ability to breed selectively for differing psychophenes and to produce large numbers of individuals with identical genotypes by systematic inbreeding. Since mammals and birds and some other groups care for their young for a protracted period, the possibility of nongenetic transmission of modes of behavior is always present. Cross fostering of young provides a means of separating genic influences from the tutelary influences of parents or parent surrogates.

The list of species and types of social behavior that have been studied genetically is long (Fuller and Thompson, 1978; Ehrman and Parsons, 1976). Here are a few examples of the findings. Inbred strains of mice differ greatly in the frequency and intensity of fighting among males; peaceable and aggressive strains have been produced by selection. Variability in the amount of aggression can produce differences in social organization. Litters of wire-haired fox terriers, a breed selected to be scrappy, become organized despotically. Litters of beagles, a breed selected for group living, develop a weak dominance hierarchy. Mating behavior has been studied most intensively in various species of *Drosophila* (fruit flies). Assortative mating, like genotypes mating with each other, can lead to the division of a population into separate mating groups. Eventually, this can result in the formation of two species. Behavior can serve to maintain genetic diversity within a population. Frequently, the mating success of a mutant male fly is enhanced when the mutant is rare in relation to wild-type males. As the mutant males increase in numbers, their attractiveness to females decreases. Thus a balance is achieved between the numbers of mutants and wild-type individuals in the population.

The elaborate courtships of fish, birds, and insects have been studied extensively and are as characteristic of a species as its external morphology. Their importance lies in their role in synchronizing the reproductive cycles of the two sexes. Observations on the mating behavior of hybrids between closely related species provide insight into the way in which the complex communication patterns are organized genetically. Platyfish (*Xiphiphorus maculatus*) and swordtails (*X. helleri*) do not normally mate in nature but will do so in the laboratory. Hybrid males court platyfish females with indifferent success, apparently because, although they display bits and pieces of the courtship display, the parts are not sequentially integrated into an effective pattern. A similar inheritance of elements of courtship ritual with loss of coordination among the parts was observed in second-generation hybrids of mallard and pintail ducks. In field crickets (*Teleogryllus oceanicus* and *T. commodus*), the same genome regulates the emission of a highly specific calling song by the male and selective receptivity by the female (for citations, see Fuller and Thompson, 1978; Ehrman and Parsons, 1976).

The genetic transmission of courtship elements without the coordination that makes them effective in communication could indicate the absence of a master gene responsible for fitting the pieces together. A more plausible hypothesis is that the independent assortment of the genes regulating the various elements produces a mix of patterns that transmit a garbled message. In this model, an element of behavior as described by an ethologist need not necessarily correspond to a single gene. Instead, it may be associated with a group of genes that have been selected as a combination because they facilitate an adaptive unit of behavior. In the case of the field crickets, it has been shown that the finest elements of the calling song are regulated by multiple genes. Likewise, a given gene may be associated with a number of behavior patterns because its product is essential for all of them. There is no necessary congruence between the units of genetics and the units of behavioral description either in terms of the action patterns of ethology or the traits of psychology.

Do these findings with animals have any relationship to humans? It is obvious that our species displays a variety of forms of social organization as great as all other primate species combined. We also communicate nonverbally but in a less stereotyped fashion than do platyfish, ducks, or crickets. Within human as well as animal groups, individuals differ greatly in aggressiveness and in sexual drive. The amount of genetic research in these areas is miniscule compared with the effort devoted to the heritability of IQ. Concerning aggression, there is a popular belief that many human problems arise from a conflict between our innate instigation to aggress when thwarted and the social dictum that we should be peaceful. In this view, the instigation to aggress was adaptive in our ancestors, for it led to success in physical competition. Now it only produces sympathetic arousal that, finding no outlet, leads to hypertension and ulcers. It would be interesting to know how great an influence one's genes have on one's tendency to be aggressive or peaceful, but we have almost no data on this matter. (Parenthetically, despite my concern with violence in American and other societies I must agree with Wilson [1975a] that our species is by no means the most violent of mammals and that it is more notable for its cooperative than for its aggressive behavior. Unfortunately, when humans are organized into large, cooperative groups that compete with other large, cooperative groups the destructiveness of intergroup conflict, when it does occur, is greatly enhanced.)

Similarly little is known of the genetics of variation in human sexual behavior, although data from certain rare inherited disorders prove that chromosomal sex, endocrine sex, and gender identification can vary independently. Some have suggested that homosexuality is heritable, but the evidence is not convincing (Heston and Shields, 1968). In technologically advanced societies, we are observing a progressive emancipation of sexual activity from procreation and more emphasis on its recreational and social binding aspects. The change may be more a matter of opening up discussion of the subject rather than an indication of fundamental changes in behavior. There are, however, implications for human evolution in the technology and social attitudes that permit individuals to be active sexually without procreation. The genes of those who follow this course for a lifetime have no future. As Wilson points out, they may maintain their inclusive fitness by aiding siblings, nieces, and nephews, but I suspect that such altruism does not compensate for childlessness.

The possibility of effective selection in a population is a

function of the variance in family size among its members. Elective contraception tends to increase this variance; hence the possibility of effective genetic selection in humans is now as great as or greater than it was in the past (Crow, 1961). This view contradicts the common belief that public health procedures have all but eliminated natural selection as a factor in human evolution. Instead, such measures have shifted the basis of selection from disease resistance to the exercise of choice in reproduction. Nevertheless, although the variance in family size that is essential for natural selection is found in humans, we shall not know whether it has genetic significance until it is demonstrated that the gene pools of small (including zero-sized) and large families are different.

Lack of information on those aspects of human behavior genetics that are most central to the application of sociobiological principles to our species renders any prediction of the future direction of human social evolution premature. In animals, the heritability of social behavior is well documented, and social organization is affected directly by the behavior of group members. I suspect that the same may be true for small human groups, such as the nuclear family, but not for groups of 100 persons, where tradition is important. In large human groups, status, behavioral traits, genotypes, and reproductive success may vary independently. To the extent that they do, the simple sociobiological model is inadequate to explain either the persistence or the disappearance of particular forms of behavior. We simply do not have the information necessary to judge how effective natural selection based on behavioral variation is in *Homo sapiens*. I am in sympathy with much of S. J. Gould's (1976a) critique of Wilson's speculations in his final chapter that cultural differences between groups may be attributable in part to genes. Tradition and the availability of natural and technical resources must be responsible for most of the cultural differences between nations, tribes, and social classes. But the evidence for a heritable component of human ability and personality is strong (Fuller and Thompson, 1978), and it is plausible that certain cultures favor the survival and propagation of particular genotypes. Certainly, however, genetic differences between races or social classes are small in comparison with differences within such groups. Furthermore, the causes within and between group differences may not be the same; genes may play a much larger role within groups. One can also reason that the absence of differences among groups can be considered as evidence for strong genetic effects. It is in this sense that I stated earlier, "It takes a human genotype to produce a human mind." Thus I do not go as far as Gould in rejecting sociobiological explanations for human behavior while apparently accepting them for insects and lower vertebrates.

Heritability and Phenostability

The heritability of individual differences is one of the problems of behavior genetics. The other, more important for our present purposes, is the degree to which a genotype established at fertilization determines the ultimate phenotype of an individual. In some cases, such as the cricket songs referred to earlier, the pattern seems to be precisely regulated by genes (Bentley and Hoy, 1974), although even here the tempo of the song varies with temperature. In *Homo sapiens* and in other species where much of the behavioral repertoire is acquired through learning, such extreme genetic determinism is not the rule. Monozygotic cotwins are commonly called *identical*, but the appellation is accurate only for their genotypes, not for their phenotypes. Similarly, individuals from a highly inbred strain of animals are almost identical genetically, yet they too vary among themselves in behavior. I do not wish to overemphasize the plasticity of behavioral development. Monozygotic cotwins are much more alike than like-sexed dizygotic cotwins. Members of an inbred strain of mice resemble each other physically and behaviorally to an amazing degree. It is risky to assume that any behavioral trait is independent of constraints imposed by genotype.

A useful way of looking at the situation is to consider that under reasonably controlled conditions a genotype determines a *norm of reaction* around which individuals of that genotype cluster. Deviations from the norm, usually defined as the mean, are attributable to random environmental factors. Inherent in this model is the assumption that large and systematic variations in the environment may shift the norm of reaction significantly. With inbred strains of fruit flies or mice, we can determine the norm of reaction very precisely. The high and low limits of the cluster around the norm define the *reaction range*. These limits can never be determined exactly, but with a large enough sample of flies or mice one can estimate the probability of any given size of deviation. I have called the region spanning two standard deviations on both sides of the mean the *reaction zone* and used its size in relation to the norm as an index of phenostability. The index is 1.0 when all individuals of a given genotype are exactly alike on the character of interest, and it approaches, but never reaches, zero as character diversity increases.

The nature-nurture issue differs depending on whether we are considering the heritability or the phenostability of a trait. Heritability is a characteristic of a population; phenostability is an estimate of the modifiability of an individual. High phenostability implies that once a genotype is fixed at fertilization some traits of that organism are rather precisely determined, and it is compatible with either high or low heritability. This point can be illustrated by comparing the phenostability of two traits in a population of inbred C57BL mice. The heritability of all traits in such a group is zero, since its members are genetically identical. Almost all of these mice will drink a large amount of 10 percent alcohol in a free choice between it and water; the trait is phenostable. Other mice tend to avoid the alcohol. In contrast, half of these mice use the left paw in reaching for food, half the right (Collins, 1968). Paw preference is phenolabile in the sense that genotype does not specify whether the right or left becomes favored for skilled tasks. Even here, the distinction between phenostability and phenolability depends on how we define the phenotype. If it is considered to be the development of a lateral preference regardless of direction, we would conclude that the trait is phenostable. Almost no mice are ambidextrous.

Let us apply this mode of analysis to human language. All humans, except the most severely retarded, communicate by language (phenostability), although their words and grammars differ (phenolability). The fact that a child transplanted from Vietnam to the United States will grow up to speak English and subsist on hamburgers and French fries is certainly proof that behavior is not encoded in genes. But the ability of that child to function in an

American society is dependent on its human genotype. Despite the publicized achievements of such human-reared chimpanzees as Kathy, Washoe, and Sarah, we scarcely expect them to attend public schools and obtain social security numbers. I do not think that this is a trivial point. To behave as a human without a human genotype is impossible. Environment must operate within the possibilities and constraints imposed by that genotype. Stimuli modulate ongoing behavior and channel development along certain pathways, but the impetus for action and for development comes from within the organism.

Sociobiology and Behavior Genetics

The message of behavior genetics to sociobiology is that the latter should focus more attention on the relationship between genotypic and psychophenic differences. Genetic and behavioral organization are mutually interactive, but they are noncongruent systems. Genes do not code for sex drive, aggression, or altruism. All that a gene can do is to direct the synthesis of molecules that become part of an organism or that regulate the organization of the structures that are used in mating, fighting, or thinking. In reading Wilson's *Sociobiology,* one might get the impression that there are altruistic and selfish genes; but there is no evidence for their existence. I doubt that sociobiologists believe in their existence, although their language sometimes makes it appear that they do. Each simple motor component that makes up the complex song pattern of crickets is regulated by multiple genes. The fine tuning of behavioral capacities in humans must likewise be accomplished by the action of many genes at many stages of development.

Despite the separation of genes and psychophenes by complex intervening mechanisms and the noncongruence of genetic units with behavioral units, I am impressed by the success of the sociobiological approach in accounting for the varieties of social organization found among animals. If one adds Trivers' (1971) reciprocal altruism to the system, much of human behavior seems to fit the model of a behavioral code shaped by the principle that actions that maximize the survival of one's genes are favored. Parents do sacrifice more for their children than for strangers or even for their own postreproductive parents. Xenophobia is as characteristic of humans as of ants, mice, or baboons.

Of course, our social mores can also be explained by nongenetic transmission. Social learning and imitation are much in vogue as mediators between the generations, and no one can doubt their role in producing cultural differences. But with regard to the core behavior of being human, such theories are no more rigorously proven than are the sociobiological ones. Learning seems to require reinforcement, and we must ask why soft words, a smile, or a touch of the hand are reinforcing. Is it because children learn that these are associated with the satisfaction of basic physiological needs for food, water, and air? Is it because they learn to fear punishment if they fail to live up to prescribed modes of behavior? Both kinds of factors may be involved, but I doubt that either is a complete explanation. Some years ago, I raised puppies in isolation from humans and other dogs during the period in which they normally become socialized. At emergence from confinement, these young dogs were fearful and withdrawn. But as they habituated to their new environment they came on call to a human experimenter, followed him as he moved about, and solicited attention when he was quiet. Play behavior with objects and with other puppies developed, less intense than with normally reared animals but still perfectly characteristic of their species. Food rewards were never employed in this delayed socialization; soft words and gentle contact were the only reinforcers. I am convinced that these reinforcers are as primary—as genetic, if you will—as hunger and thirst, even though we cannot define them in terms of the reduction of peripheral physiological imbalances.

In appraising the applicability of sociobiological principles to human beings, we must evaluate the degree to which genetic constraints operate on the plasticity of human development. Does our evolutionary history as hunter-gatherers, herders, and agricultural villagers incapacitate us psychologically for life in cities? Is our predilection for suburbia, exurbia, and summer cottages an indication of an innate need for stimuli that are more natural than steel, glass, and concrete? More fundamentally, are social changes centering about the increasingly unstable nuclear family running counter to psychological needs that are likewise a part of our basic, genetically influenced, human nature? Do children have a need to identify with parent figures in order to manifest their full human potential?

Sociobiology espouses a conservative point of view: Our genotypes and cultures are the products of evolution and natural selection, and they are fitted to each other. There is, in Wilson's terminology, a phylogenetic inertia that slows down adaptation to new conditions. Thus we share with other primates tendencies for territorial defense, for mistrust of strange conspecifics, for male dominance, and for a prolonged period of socialization of the young. Presumably these patterns developed because they were adaptive during the evolution of *Homo sapiens*. Can we now move on to a new kind of society in which some of these characteristics will be eliminated? Can they all be eliminated? Are the barriers to such change merely those of eliminating habits or are there biological changes that must occur? Currently in the Western world there is a challenge to the traditional difference in the social roles of males and females. Sexual differentiation is clearly genetic, and the physiological and structural correlates are obvious. I believe that the widening of role choices for both sexes is a great advance but confess to doubts that the differences between the interests of men and women are solely the product of exposure to sex stereotypes in the nursery and in their school readers. Time will tell. Testosterone and estrogens are potent compounds that affect the brain as well as the reproductive system.

What of our propensities for establishing territories, for dividing into social classes, for resorting to violence between races and nations? Parallels to all these exist in other primates. Within a primate band, overt aggression is minimized by the development of hierarchical dominance systems. Humans also have dominance systems that are at the same time more elaborate, more formal, and more flexible than those of our primate relatives. The question of whether aggressive, dominant, and submissive modes of behavior are innate or learned is an empty one. Aggression seems to be elicitable from any normal human except possibly one trained explicitly to be submissive or one subjected to a history of repeated defeats. Sociobiology asserts that we can respond aggressively today because the ability to fight enhanced the fitness of our ancestors; it does not claim that a need to aggress is carried in the genes. Sociobiology also views hierarchical social organization as adaptive because it reduces intragroup conflict. To some, hierarchies have an elitist tinge; the strong take unfair advantage of the weak. But any species that lives communally has two choices: It either accepts organization with some centralization of power, or it engages in perpetual conflict for scarce resources. Fortunately, the choices are not restricted to the extremes of anarchy or despotism. Humans have taken a middle course with a multidimensional social organization. It is a course that still produces more inequity and more violence than we like, but we cannot blame sociobiology for problems that antedate it by millennia. There is no basic conflict between sociobiology and the behaviorist position that violence increases if rewarded and declines if not rewarded.

Large-scale intergroup aggression is a complex social phenomenon that must be considered very briefly. Durham (1976) has proposed that intergroup aggression develops as a joint product of biological and cultural evolution. The prevalence of tribal warfare in many areas negates the notion that it is a product of nation-states or high-level technology. Any sociobiological explanation of this almost exclusively human trait must postulate that participation in war enhances the fitness of its participants despite the risks of death or injury. Can we explain differences among cultures in their predilection for war by differences in their genes? I doubt it. It is almost certain that all human groups are descended from ancestors who fought intruders on their territory. It is likely that the genetic capacity for violence is latent in every human being, just as is the latent capacity for submission. The unique quality of the human genotype is that it has guided the development of a brain with many options, each as natural as any other. Whether it has any choice in the exercise of these options—that is, whether the brain is the seat of free will—is a philosophical issue I shall avoid for the present. Behavior genetics suggests that the brain has preset biases to learn certain modes of behavior, oral speech more readily than reading, and aggressive rather than conciliatory responses to frustration. We have the ability to design societies that will favor the expression of those qualities we value the most.

Conclusion

I do not know how genes produce a human brain and mind. I do know that in our species, and in all others that have been studied adequately, genetic variation in behavioral capacities exists. The great difference between humans and other species is that our brain extends the range of its storage capacity and decision-making processes beyond immediately present stimuli to those encountered in the past and visualized for the future. Our brain is the only one that looks into itself and tries to understand its own functioning. That the genome that guides the development of this brain is the product of natural selection seems obvious to me, but saying this does not imply genetic determinism of behavior. Rather, it signifies a genetic basis for a programmable integrating mechanism connected to sense organs, glands, and muscles.

Since human genetic variation is so prevalent, it seems inconceivable that our species is not still subject to natural selection and future evolution, if it does not destroy itself in a nuclear holocaust or poison itself with chemicals. Even these events might be construed as a drastic form of natural selection. In my darker moods, I think that humans might have done better to have stayed a little closer to other primates, who seem to enjoy themselves immensely in their natural habitats but whose brains are inadequate to develop a technology of atomic weapons and pesticides. But such thoughts are fleeting—how exciting it is to have these brains that can press so far beyond the concerns of our animal cousins, brains that can communicate across cultures and between remote generations!

Does the sociobiological view help us to understand ourselves? I think so, even though it is pretentious to consider sociobiology as giving ultimate explanations. I would prefer to think of it as a kind of historical reconstruction pieced together from fossils, observations on contemporary species and a plausible, but untested, genetic model. History can be enlightening and may even help us to predict the future, but functional explanations of human behavior based on observation and experiment in the contemporary world are more useful guides for the amelioration of inequities and hazards in our societies. Sociobiology will affect our efforts in these directions by its influence on our philosophical views of human nature and on our hierarchy of ethical values.

6

Frank A. Beach

Sociobiology and Interspecific Comparisons of Behavior

I begin this chapter under a double handicap. First of all, since I am a comparative psychologist anything I say may seem no more than the ghostly echo of a voice from the grave, because the high priest of sociobiology, E. O. Wilson, has publicly prophesied the death of my discipline, proclaiming it "destined to be cannibalized by neurophysiology and sensory physiology from one end and sociobiology from the other" (1975a, p. 6). Wilson is able to contemplate this grisly feast with equanimity because of its purgative results:

> The future, it seems clear, cannot be with the ad hoc terminology, crude models, and curve fitting that characterize most

The author's research mentioned in this article was supported by USPHS Grant 04000 from the National Institute of Mental Health.

of contemporary ethology and comparative psychology. Whole patterns of animal behavior will be inevitably explained within the framework, first of integrative neurophysiology . . . and, second, of sensory physiology. . . . To pass from this level and reach the next really distinct discipline, we must travel all the way up to the society and the population. Not only are the phenomena best described by families of models different from those of cellular and molecular biology, but the explanations become largely evolutionary. . . . As Lewontin has truly said, "Natural selection of the character states themselves is the essence of Darwinism. All else is molecular biology."

I can only admire and envy the nonchalance and intrepidity with which Lewontin fearlessly leaps from the heights of molecular biology to the even loftier peak of "character states," whatever they may be; but as a journeyman scientist I am more concretely concerned with identifying intermediate stages along the route behavioral science is to follow in progressing from integrative neurophysiology, "all the way up to the society and the population" (Wilson, 1975a, p. 6).

This is my second handicap. I recognize that in the postulational-deductive approach of sociobiology the postulates are sacrosanct by definition and need never be proved. I also accept the proposition that postulates can be used in building models intended to illuminate general principles. However, a distressing myopia impedes my labored search for reassuringly tangible connections between, on the one hand, the sociobiologist's formidable postulates and impeccably mathematical theoretical models and, on the other hand, the grubby raw material of empirical evidence with which a pedestrian comparative psychologist perforce must deal. I am haunted by the conviction that somewhere, somehow, there should and must be a connection, but it perversely eludes me.

Part of my difficulty may stem from a naive faith in the proposition that the behavior of the individual must be understood in terms of what sociobiology calls "promixate" causes. This really is a basic enterprise as far as psychology is concerned. Its value is not denied by sociobiologists, but the discovery of "ultimate" causes is valued much more highly, and it is their elucidation that inspires the proclamation of postulates.

A second obstacle to my recognition of essential connections between some sociobiological models and an empirical data base

is the model makers' apparent omission or disregard of facts concerning interspecific similarities and especially interspecific differences. The position of most surviving comparative psychologists is
that behavioral science begins with analysis of individuals, proceeds
to generalizations encompassing the species, compares and con-
trasts species, and hopes for discovery of broader principles with
greater or lesser interspecific validity. It appears to me that some
exponents of sociobiology are concerned with empirical observation only insofar as it serves a particular theoretical end. Furthermore, there is at least the suspicion that the validity or reliability
of behavioral evidence may be less rigorously evaluated than its
compatibility with the original model.

Two Kinds of Criticisms

To conclude these already personal introductory remarks,
it is appropriate to say explicitly that I see much that is valuable,
original, and even exciting in current sociobiology. For example,
recent field studies of social behavior of the Olympic marmot
(Barash, 1973a) and Belding's ground squirrel (Sherman, 1977)
are nonpareil, and much of their value stems from sociobiological
theory but is grist to the mill of comparative psychology. I am neither so confident nor so arrogant as to propose any sort of indict-
ment of an entire field or of any contributor thereto. Criticisms
raised in the following pages arise from two sources. One may well
be my own failure to understand fully the true aims and rationale
of sociobiology. The other is a sincere conviction that, whatever its
virtues and no matter how great its promise, sociobiology has several weaknesses and blind spots. To illustrate both categories, I will
discuss very briefly the apparent reliance on unreliable evidence
and then deal with what I consider a genuinely fundamental issue,
namely the problems and principles involved in comparing presumably similar behavior of different species.

Evaluation of Empirical Evidence. Various critics of sociobiology have complained, and I agree, that evidence harmonious with
basic postulates sometimes is used for model building with little or
no examination of the reliability of the original observations or
validity of the reporter's interpretation. A case in point is the oft-

cited "infanticidal" behavior of male langur monkeys (Sugiyama, 1967; Mohnot, 1971; Hrdy, 1974, 1977).

As usually described, this phenomenon begins when a heterosexual troop of langurs (*Presbytis entellus*) is invaded by one or more adult males who succeed in displacing the original dominant male. During the accompanying fighting, which may go on for days, a new male assumes leadership and subsequently kills those young that are still of nursing age. Wilson (1975a, p. 85) puts it more colorfully by reporting that "the young are actually murdered by the usurper."

In terms of prevailing theory, this infanticidal behavior has at least two important functions. First, it eliminates infants that do not carry the genes of the new male, so that he need invest no energy in contributing to the survival of unrelated individuals. Second, lactating females deprived of their young soon come into estrus and are then available for impregnation by the new leader, who can thus increase the perpetuation of his own genotype.

Difficulties encountered by this interpretation arise in large measure from disagreement as to the significance and especially the "representativeness" of the original behavioral observations. No one denies that in specific situations at least a few male langurs have inflicted fatal wounds on infants of nursing age. The central question is whether infanticide truly reflects an evolved "reproductive strategy" (Hrdy, 1977) or occurs as a consequence of social stress arising from unnatural living conditions (Curtin, 1977; Curtin and Dolhinow, in press). Authorities espousing the second alternative point out that infant killing has been reported for only a few troops of langurs that were living in close proximity to human habitation and that comparable behavior has not been seen in social groups inhabiting other environments. The point at issue would seem to be whether infanticide is a behavior pattern typical of the species or is simply an infrequent, aberrant, and extraneously induced event. Sociobiologists may reply that the distinction is irrelevant because it is entirely legitimate to build models on exceptional cases or even on no empirical evidence at all; but this position is weakened when the model is seen by critics as an attempt to endow the original observation with special functional significance.

Of course, the value and importance of the sociobiological approach are not to be judged in terms of isolated examples of debatable evidence, but leading expositors of any theoretical position are well advised to select for purposes of illustration behavioral descriptions whose reliability and representativeness are generally accepted.

Interpretation of Behavior and Interspecific Generalization. A different desideratum for model making pertains to the validity with which observed behavior is interpreted, and this becomes particularly important when more than one species is involved. It is in connection with interspecific comparisons that sociobiology differs markedly from comparative psychology. The nature of the difference can be illustrated by the ways in which these disciplines deal with homosexual behavior in animals and in humans.

Homosexuality is mentioned a number of times in *Sociobiology* (Wilson, 1975a) but in addressing his readers Wilson seems to rely on the implicit assumption of a mutual understanding, for at no point is the key term given explicit definition. It is stated, for example, that female rhesus monkeys show "aberrant" responses, such as homosexuality and masturbation, when the structure of the social group is disrupted (Wilson, 1975a, p. 22). On the same page, socially subordinate male South American leaf fish (*Polycentrus schomburgkii*) are said to exhibit homosexuality when they temporarily assume body color patterns typical of a female in what Wilson describes as "an attempt to fool the resident males and to 'steal' a fertilization by depositing their own sperm around the newly laid eggs" (p. 22). This second usage of the term *homosexual* becomes even more confusing when Wilson concludes that if his interpretation is correct it represents "a case of transvestism evolved to serve heterosexuality" (p. 22). The effortless transition from homosexuality to transvestism would surely dismay serious students of human sexuality, who have gathered a great deal of evidence establishing the independence of the two phenomena in our own species.

In cases such as this—and their number is legion—disingenuous claims that no anthropomorphism is intended do not protect sociobiology against the charge of linguistic irresponsibility. If words from everyday language are intended to convey an extraordinary meaning, the specialized meaning must be explicitly defined. Complete disregard for this elementary rule has particularly damaging consequences when the same term is applied indiscriminately to fishes, monkeys, and humans without any acknowledgment of species differences in causal factors or functional outcomes. The fact that such comparisons may serve to illustrate a favored postulate does not negate their inadmissability to scientific reasoning.

Instead of purely nominal comparisons coupled with egregious interspecific extrapolations, what is called for is systematic, species-by-species analysis of *normal behavior patterns*. When this has been accomplished, we *may* have the foundation for fruitful postulates of general evolutionary significance. I emphasize the phrase "normal behavior patterns" because nonspecialists unfamiliar with the analysis of sexual behavior are apt to focus immediately on that which their own culture designates as "abnormal," and this approach is counterproductive for three reasons. First, we cannot even define "abnormal" behavior until we have clearly identified the limits of normality. Second, when the latter is accomplished we may well discover that responses originally designated as abnormal are not that at all. Third, even behavior that falls more than 3 sigma from the mean will defy explanation until we have arrived at a satisfactory understanding of the basic pattern from which deviation has occurred.

Homosexual Behavior in Animals

As applied to nonhuman mammals, the term *sexual behavior* usually refers to activities involved in the fertilization of ova; and in mammals this consists primarily of the male and female copulatory patterns.

Types of Interactions Observed. Laypersons and scientists who have never studied the mating behavior of animals commonly assume that it is exclusively dimorphic and heterosexual, but the truth is that the relations involved are less rigidly dichotomous than is generally recognized. Each of the six variants listed in Table 1 occurs spontaneously—that is, without experimental intervention—in one or more species of mammal. For example, Type 3 is not at all uncommon in dogs and monkeys, and Types 5 and 6 together have been recorded for thirteen species of mammals representing five different orders (Beach, 1968).

As indicated in the table, some investigators have applied the term *homosexual* to male and female animals in two different circumstances. First, it has been used to describe individuals that exhibit coital reactions typical of their genetic sex but that do so in response to a like-sexed partner. This is applicable to the mounting behavior of males in Types 3 and 4 and to receptive reactions of recipient females in Type 6. In such cases, it could be argued that the critical feature is simply a perceptual error or lack of discrimination resulting in failure to detect the sex of the partner.

A second usage of the term *homosexual* is reflected in its application to animals that exhibit the coital pattern typical of the opposite sex. Such is the case with respect to recipient males in Type 4 and to initiating females in Types 5 and 6. Behavior of this sort is more correctly classified as *inversion of motor patterns.* The distinction was pointed out thirty years ago by Kinsey, Pomeroy, and Martin in their classic study, *Sexual Behavior in the Human Male* (1948, p. 614–italics added):

> The assumption by a male animal of a female position in sexual relation or the assumption by a female of a position which is more typical of the male in a heterosexual relation is what the students of animal behavior have referred to as "homosexuality." *This, of course, has nothing whatsoever to do with the use of the term among students of human behavior, and one must be exceedingly careful how one transfers the conclusions based on these animal studies.* In studies of [human] behavior, the term "inversion" is applied to sexual situations in which males play female roles and females play male roles in sex relations. Most of the data on "homosexuality" in the animal studies actually refer to inversion.

When terms designating complex and poorly understood aspects of human behavior are indiscriminately applied to unanalyzed responses of other species, ignorance is compounded, and confusion reigns. The way to understanding is not further juggling of definitions, but more precise analysis of the original evidence— that is, of directly observable behavior.

Table 1. Types of Copulatory and Quasi-Copulatory Relationships Observed in Some Species of Mammals

| | Initiator of Mounting Activity | | Recipient of Mounting Activity | | |
Type	Genetic Sex	Popular Designation	Genetic Sex	Behavioral Response	Popular Designation
1	Male	Normal	Female	Resistant or passive	Normal (unreceptive)
2	Male	Normal	Female	Cooperative[a]	Normal (receptive)
3	Male	Homosexual	Male	Resistant or passive[a]	Normal (or homosexual)
4	Male	Homosexual	Male	Cooperative	Homosexual
5	Female	Homosexual	Female	Resistant or passive	Normal
6	Female	Homosexual	Female	Cooperative	Homosexual

[a]The term *cooperative* implies full display of the feminine pattern of receptivity regardless of the sex of the responding individual. Sometimes the importance of a positive response is overlooked, so that a passive individual that is mounted by another of the same sex and fails to exhibit active resistance is labeled "homosexual."

Variables Involved in Normal Mating. Sexual behavior of animals is analyzable in terms of three independent variables. The first is the genetic sex (XX or XY) of each participant.

The second is the "sex" of the behavior pattern displayed. These directly observable responses consist exclusively of copulatory and precopulatory reactions. For nearly all mammals, the masculine pattern includes mounting, pelvic thrusting, penile insertion, and ejaculation. The feminine pattern comprises assumption and maintenance of a posture that facilitates the male's achievement of insertion and intravaginal ejaculation. This is termed *receptive behavior,* and its execution may be preceded by, or interspersed with, various invitational or solicitational activities collectively defined as *proceptive behavior.*

The male genotype and the masculine mating pattern are not exclusively correlated, nor do normal females exhibit only feminine copulatory reactions. It is an obvious but important fact that both sexes are entirely capable of executing many of the coital responses characteristic of the opposite sex. When genetic sex and overt pattern are congruent, the behavior is described as *homologous.* If they are incongruent, as when one female mounts and thrusts on another female, the behavior is *heterologous.*

The third variable to be accounted for is the "sex" of the stimulus pattern to which an individual is reacting. The feminine stimulus pattern comprises characteristic odors, possibly certain visual signs present only in the estrous female, and the receptive and proceptive activities typical of females when they are ready to mate. The masculine stimulus pattern may involve typical male odors, visual characteristics, special calls, and particularly the tactile and pressure stimuli applied by the male as he mounts and penetrates the female.

When mating behavior of different species is compared in terms of these three sets of variables, certain principles or first-order generalizations become evident. First is the *principle of S-R complementarity,* which states simply that the feminine pattern of sexual stimuli has the highest probability of eliciting the masculine pattern of mating reactions and, conversely, that the masculine stimulus pattern tends to evoke feminine copulatory and precopulatory responses. These relationships would be unremarkable were it not for the fact that they hold true *regardless of the sex of the individuals involved.*

The principle of S-R complementarity is thus independent of genetic sex, which means that when exposed to the feminine stimulus pattern both males and females, if they respond at all, usually exhibit masculine coital reactions. In contrast, for animals of either sex the masculine configuration of stimuli is most likely to elicit feminine mating responses.

If S-R complementarity were the only determinant of sexual interactions, the result would be a condition of balanced behavioral bisexuality, but this never occurs in mammals because of the bias introduced by a second principle. This is the *principle of sex-linked prepotency,* and it applies equally to motor patterns and to stimulus sensitivity. In both sexes, homologous behavior (that is, behavior congruent with genetic sex) is prepotent (more easily elicited) over heterologous reactions. In complementary fashion, both males and females are more sensitive to heterologous than to homologous patterns of sexual stimulation.

Male-Male Relations. Under certain specifiable conditions, sexually aroused male mammals of numerous species exhibit mounting responses indiscriminately directed to conspecifics of either sex or even to inanimate objects. Collection of semen to be used in artificial insemination is achieved by interposition of an artificial vagina while a specially conditioned male is mounting and thrusting on another male, in the case of bulls, or on a wooden dummy, in that of stallions. Rabbit bucks will simulate copulation and ejaculate when confronted with the fur-gloved hand of an attendant. Regardless of species, a particularly effective method of eliciting the complete masculine pattern, including ejaculation, involves preexposure of the male to a conspecific female in heat.

Two conclusions are indicated. First, parameters of the perceptual pattern capable of evoking mounting responses are not rigidly confined to specific sensory qualities of the estrous female. Second, despite this obvious latitude the principle of S-R complementarity is not violated, because the effectiveness of "reproductively inadequate" stimuli depends on previous conditioning and preceding exposure to the normal heterologous pattern.

The second conclusion receives additional support from

studies of male-male interactions that occur when one male displays mating responses characteristic of a sexually receptive female. Spontaneous display of the feminine pattern by untreated males is uncommon in nonprimate mammals, although its frequency varies among species. An especially significant point is that individuals exhibiting this type of inversion are fully responsive to the heterologous stimulus pattern and copulate vigorously with receptive females when given the opportunity to do so. The overall behavior repertoire is therefore correctly designated as *bisexual*.

The relevance of such a repertoire to the complementarity principle is twofold. First, when one male rat, for example, is confronted with a second male that displays heterologous mating responses, the mounting behavior of the first animal is greatly intensified. A passive or a slightly resistant masculine partner may elicit some mounting and thrusting reactions, but a second male behaving like a female will call forth much more frequent and vigorous coital behavior. Second, if a "bisexual" male is tested first with a partner of his own sex and then with a receptive female the male will show the feminine response pattern while being mounted by another male and will shift immediately to the masculine pattern of mounting and ejaculating as soon as the estrous female becomes available (Beach, 1945). In this clear-cut example of S-R complementarity, the "sex" of the motor pattern is completely determined by the "sex" of the stimulus pattern, so that sex-linked prepotency is apparently lacking.

The occurrence of coital responses in nonprimate mammals is strongly influenced by gonadal hormones. Female rodents, carnivores, and ungulates usually exhibit receptive behavior only under the influence of estrogen or estrogen plus progesterone. Males occasionally display mounting responses in the absence of gonadal secretions, but testosterone is required for normal performance of the complete masculine repertoire. Male rats that have been castrated at birth before sexual differentiation is complete react to injections of estrogen in adulthood by displaying the entire feminine copulatory pattern. When the same animals are treated with testosterone, they copulate essentially as normal males except for the absence of ejaculation, due to extreme smallness of the penis, which greatly reduces the frequency of intromission.

In various species of monkeys and in chimpanzees, it is common to observe a form of male-male interaction in which one individual "presents" to another with back turned and hindquarters exposed by bending forward and in which the second male responds by grasping and briefly mounting the presenting animal. The functional significance of such behavior has been variously interpreted and may in fact differ according to the situation. Most commonly it has been taken as an expression of dominant-subordinate relations between the two males, but it may also constitute a gesture of friendship or comradeship. In such cases, no sexual relevance need be assumed. On the other hand, there is at least one report of homosexual anal intercourse with ejaculation in the rhesus macaque, so it is advisable to avoid any final generalization until the evidence is more nearly complete.

There is no question that presentation is a basic component in the normal female's mating pattern. Assumption of this position facilitates the male's achievement of insertion, and the response is characteristic of sexual receptivity in many species of monkeys and apes. Therefore, when one male presents and is mounted by a second male the total interaction is consonant with the principle of S-R complementarity.

Female-Female Relations. In a number of species, female-female mounting is much more common than comparable behavior between males. Females that respond to others of their sex with the masculine coital pattern are nevertheless normal in their execution of homologous (feminine) mating reactions when paired with male partners during estrus.

Mounting and thrusting by female mammals is most vigorous and frequent when the mounting individual is exposed to another female that is in heat and therefore presents the full feminine stimulus pattern. Thus the principle of S-R complementarity is preserved, despite the genetic femaleness of both partners. I consider it unnecessary to define receptive performance of the stimulus female as "homosexual," since she is responding in normal feminine fashion to the heterologous stimulus pattern. In fact, she will react in the same manner if the experimenter stimulates her flanks and perineum with his fingers.

Gonadal hormones contribute to the normal occurrence of

interfemale mounting in two ways (Beach, 1968). First, they are essential for display of sexually receptive behavior (the homologous pattern), and therefore they increase a female's capacity to stimulate or elicit mounting by another animal of her own sex. Second, the tendency for females to exhibit mounting responses is intensified by the same hormones that produce receptive behavior. In fact, females of most species are most likely to exhibit mounting behavior when they themselves are in heat and therefore are sexually receptive. For some species, such as the guinea pig, female mounting is a normal accompaniment of estrus and occurs in no other phase of the reproductive cycle. Among female dogs, swine, sheep, cattle, and horses, a sudden onset of mounting activity signals that the female is or shortly will be in heat, and the behavior is in no sense "abnormal."

In rodents, dogs, and monkeys, this normal tendency to mount may be intensified if females are treated with androgen; and the effect is magnified if testosterone is administered twice, first during prenatal or neonatal sexual differentiation and second in adulthood when behavior is tested. However, as noted earlier, exogenous androgen is not necessary for the occurrence of the heterologous coital pattern.

Female chimpanzees and female monkeys of several species often present to one another, and under some circumstances the recipient clasps and briefly mounts the presenting individual. Like comparable behavior in males, this type of interaction has often been characterized as a manifestation of social dominance and submission. In most cases, any relation to coital behavior remains highly speculative, and current evidence certainly does not justify a definite diagnosis of "homosexuality." Here again, however, there may be exceptions, for at least one observer has described instances of female-female mounting considered frankly sexual because of the inferred occurrence of climax in the mounting individual (Chevalier-Skolnikoff, 1974).

Interpretation of Observations on Animals. Evidence presently available is too limited in quantity and uneven in quality to validate or disprove elaborate theories concerning either the development or the display of sexual behavior in animals. However, it is not premature to suggest two working hypotheses as long as they are understood to apply only to those species that have been systematically examined.

Brain Organization. The first hypothesis pertains to organization of putative brain mechanisms involved in the mediation of mating behavior. Sexual differentiation of parts of the reproductive system such as the genital organs is mutually exclusive; that is, an individual can possess a penis or a vagina but not both. This is so because both structures develop from the same original embryological tissues. For some mammals, a different situation apparently obtains in development of the central nervous system. Differentiation of neural circuits for mediation of male mating behavior does not necessitate suppression of those controlling female behavior, and vice versa. Accordingly, the same individual may possess both types of brain mechanisms.

If it is correct, this concept of neural bisexuality leads to the prediction that for any species to which the model applies, the sexual tendencies and capacities of individuals cannot be adequately conceptualized in terms of a bipolar, unidimensional scale extending from "female" at one end to "male" at the other. The schema of a single masculine-to-feminine continuum would be incongruent with the coexistence of male and female systems in the central nervous system. A bisexually organized brain could mediate behavioral expression of varying degrees of masculinity and femininity by the same individual. In terms that I have used to describe mating behavior, the notion of bisexuality of central nervous mechanisms implies that brains of genetic females and of genetic males include neural circuits capable of transducing heterologous input into homologous output and, contrariwise, of mediating heterologous responses to homologous stimuli.

Within a single individual, male and female brain systems are not equally responsive, partly because of genetically or developmentally determined sex-linked prepotency and partly because of differences in the sensitivity of these systems to gonadal hormones, as discussed in the following section.

Sex Hormones. The second hypothesis has to do with effects of sex hormones on mating behavior. In many species, the heterosexual responsiveness and potency of males are profoundly affected by androgen, just as proceptivity and receptivity of females are dependent on ovarian hormones. Furthermore, in both sexes administration of heterologous gonadal hormones tends to increase the display of heterologous copulatory behavior. There is, however, a sex difference in responsiveness such that males react less strongly than females to estrogen or to estrogen and progesterone and that females are less responsive than males to testosterone.

The second hypothesis is, therefore, as follows. In both sexes, brain mechanisms for male behavior are stimulated by androgen, and those for female behavior are stimulated by estrogen and progesterone; but the male brain reacts more strongly than the female brain to androgen, whereas the female brain is more responsive than the male brain to ovarian hormones. This sexual bias in brain sensitivity to gonadal hormones is developmentally determined, depending in part on the presence or absence of androgen during differentiation and in part on as yet unspecifiable genetic influences.

Conclusions. The temporary and reversible inversion of mating roles observed in nonhuman mammals reflects a bisexuality of brain organization that is normal for these species. This feature of the central nervous system permits both sexes to "transduce" the external feminine stimulus pattern into the motor pattern of masculine copulatory responses or to react to the masculine pattern of stimulation with execution of feminine coital responses.

Role inversion reflects the automatic functioning of preorganized S-R connections between sensory and motor systems and involves no physiological or behavioral "abnormality." However, the same forces that decree a bisexual neural basis for mating behavior also impose a sexual bias on development, such that masculine responses will be more readily and fully evocable in males than in females, whereas feminine reactions will be superordinate in females.

The effects of sex hormones on inversion of coital roles are permissive rather than obligatory. Androgen increases responsiveness to the feminine stimulus pattern in both sexes, although the effect is stronger in males. Ovarian hormones potentiate feminine reactions in females and males, but females are more responsive than males.

A special sort of control seems to be exerted by testosterone acting during prenatal and/or neonatal development. Under normal circumstances, the testicular hormone appears to influence the brain so as to reduce permanently the capacity for feminine behavior while enhancing that for masculine behavior. This is not, of course, an all-or-none effect but is probably related to the sex-linked bias toward homologous mating responses in adulthood.

Human-Animal Comparisons

What relevance, if any, has the foregoing analysis to the phenomenon of human homosexuality? This question is much more complex than most laypersons and many scientists, including psychologists and sociobiologists, seem to realize. In fact, it cannot be answered at all unless both human and animal behavior have been analyzed in terms of comparable parameters. There is a fundamental rule that applies to all such cases whether the comparison is between animals and humans or between different species of animals. *The validity of interspecific comparison is limited by the reliability of intraspecific analysis.* Meaningful comparisons between Species A and Species B simply are not possible until the behavior in question has been analyzed with equal care, objectivity, and precision *in both species.*

In the present instance, it is fruitless and potentially dangerous to attempt to compare a selected aspect of human and animal behavior when the sole basis for selection is a common label referring to superficial similarities. It is worth repeating that the essential first step is separate analysis of the behavior *in each species independently.* Only after this is accomplished can we hope to discern fundamental similarities and differences. What is fundamental is not the formal properties of the behavior, but rather its causal mechanisms and functional significance for the individual and for the species.

As far as animal sexual behavior is concerned, I have dealt with such variables as specific sensory stimuli and clearly discriminable motor patterns; but homosexual activities of men or women have never been analyzed in such terms, and it is doubtful if the results of such a procedure would be particularly illuminating. For

human beings, there are no exclusively masculine or feminine patterns of copulation. As far as homosexuality is concerned, the old
distinctions between passive and active sexual roles or between
"inserter" and "insertee" no longer are considered meaningful dimensions of a homosexual relation (Hoffman, 1977).

    With respect to sexually distinct stimulus patterns, we can
only say that their potential involvement in human homosexuality
is obscure. It is certain that the principle of S-R complementarity
does not apply. Contrary to the popular stereotype, most homosexual men are more responsive to masculine than to effeminate
partners, and lesbians are drawn to each other on the basis of mutual femininity (Hoffman, 1977). The defining characteristic of
homosexuality is its homophilic foundation rather than the overt
behavior through which homosexual attraction is manifested. Our
current inability to understand this phenomenon is unremarkable
in view of our equally profound ignorance regarding the factors
responsible for a positive sexual valence between individuals of the
opposite sex. It is entirely possible that the most effective approach
to interpreting homosexuality will prove to be the achievement of
a better understanding of heterosexual attraction and love.

    I suspect that organization of the human brain is at least
potentially bisexual and that sex hormones probably increase erotic
responsiveness in men and women, but this is just a "thunch"—
half theory and half hunch—and before thunches become hypotheses they demand verification through direct study, in this
case, study of human subjects. Certainly human homosexuality
differs from role inversion in animal mating, both in terms of
causal mechanisms and of functional outcomes.

Relevance to Sociobiology

    What bearing has the foregoing discussion on central issues
of sociobiology? The answer depends in part on the particular issue
involved.

    Validity and Representativeness of Behavioral Evidence. If one is
seeking heuristic examples to illustrate a particular hypothesis, it
may be sufficient to know that on a single occasion an adult male
langur invaded a heterosexual group, displaced the dominant
male, killed all nursing infants, and then copulated with their

mothers. From such a starting point, it is perfectly feasible to construct an elegant model that will predict whether or not, or under what circumstances, such infanticidal behavior would influence the "fitness" of the male by increasing the probable perpetuation of his own genes. This is an essentially *aspecific* approach in that the species is irrelevant and it matters little whether one is dealing with langurs or unicorns.

The model-building sociobiologist may be unconcerned by a primatologist's criticisms to the effect that some reports of infanticide are of dubious reliability or that at most infanticide can be considered a rare form of behavior associated with abnormal ecological conditions. Such complaints may be seen as irrelevant by a theorist who is concerned neither with understanding the behavior of langurs as a species nor with analyzing the proximate causation of a particular behavioral incident. This attitude obfuscates effective communication with the comparative psychologist whose principal goals are to describe, measure, and compare analogous behavior patterns in different species and to analyze behavior in terms of its motivational and mediational components. To such an individual, the sociobiologist may fit in the category described by B. F. Skinner (1938, p. 44) as "men whose curiosity about nature is less than their curiosity about the accuracy of their guesses."

My own attempts to evaluate critically the logic of the sociobiological approach are to some extent hampered by this seemingly cavalier attitude toward the pedestrian facts of everyday behavior, but I am more profoundly alienated by the logical and linguistic confusion arising from irresponsibly uncritical usage of key terms and concepts. In this connection, the foregoing discussion of homosexuality clearly is relevant.

Definition and Interspecific Transfer of Key Terms or Concepts. I have already mentioned E. O. Wilson's varied references to homosexuality in *Sociobiology: A Modern Synthesis* (1975a). The term is applied to fish, monkeys, and humans without distinction and without definition. In different instances, it refers to changing body color patterns (South American leaf fish), "aberrant" female-female responses (rhesus monkey), "ritualized aggression" (male baboons), and unspecified relations between women or between men. I may have overlooked significant passages, but nowhere can I discover precisely what the words *homosexual* or *homosexuality* are intended to mean, and this implicit assumption of a mutual understanding strikes me as dangerously naive. Neither can I find explicit recognition of the fact that proximate causes and functions of color change in fish and homosexual attraction among humans are simply incomparable.

It is all very well to point out that such differences are as obvious to the sociobiologist as they are to me and that his or her business is to unravel the tangled skein of ultimate rather than proximate relations, but this disclaimer will not undo the mischief caused by initial carelessness in the use of original terms. Profound postulates may yield exciting hypotheses that delight the heart, but, even though they need never be proved, postulates do not derive from divine inspiration. They have their earthly roots, and when these sources turn out to be unreliable descriptions of behavior, egregious extrapolations from multispecific comparisons, or simply sheer speculation, the nonspecialist's faith in the sociobiological enterprise is badly shaken.

Imputation of Genetic Correlates. My interpretation of inversion of copulatory roles in nonhuman animals assumes the existence of brain mechanisms whose organization is determined by developmental processes under genetic control. Most if not all comparative psychologists would agree that the genotype exerts important effects on behavior, but they might well cavil at the sociobiologist's simplistic inferences regarding a genetic basis for human homosexuality.

This notion surfaces again and again in Wilson's book, where implicit assumptions are combined with speculative interpretation. The first of the implied assumptions, as I read them, is that there is a unitary type of behavior that occurs in different species and that justifies a common label of "homosexuality." The second assumption is that homosexuality has, or at least might have, a genetic basis. Of course, it is emphasized that nongenetic factors must always be involved and that presence of the genetic ones does not necessitate their phenotypic expression. Finally comes speculation as to how "homosexual genes," if they existed, might function as an ultimate cause by actually increasing the "fitness" of their bearers.

The reasoning involved is slightly convoluted, but its nature is best understood in terms of specific examples taken from *Sociobiology* (Wilson, 1975a). Here is a direct quotation referring to human prehistory: "The homosexual members of primitive societies may have functioned as helpers. . . . Freed from the special obligations of parental duties, they could have operated with special efficiency in assisting close relatives. Genes favoring homosexuality could then be sustained at a high equilibrium level by kin selection alone" (p. 555). Another example suggests that homosexual genes may indeed have been sustained to the present day, when they are capable of affecting the choice of various life-styles: "The celibate monk, the maiden aunt, or the homosexual need not suffer genetically. In certain societies, their behavior can redound to improved fitness of parents, siblings, and other relatives to an extent that selects for the genes that predisposed them to enter their way of life" (p. 343).

Finally, Wilson (1975a, p. 555) quotes with apparent approval G. E. Hutchison's suggestion that "homosexual genes may possess superior fitness in heterozygous conditions." This is heady stuff indeed! We are invited not only to contemplate the potentially adaptive value of genes promoting homosexuality but also to consider the further possibility that the genes are recessive. Alert psychoanalysts might justifiably suggest that latent homosexuality represents the phenotypic expression of a "homosexual gene" combined with its dominant heterosexual allele!

Such flippancy is probably undeserved and surely unprofessional, but the overall reaction of this comparative psychologist to certain salient characteristics of sociobiology resembles the sentiments of a famous seventeenth-century naturalist and physician regarding contemporary speculations concerning the nature of magnetic force. In a letter to a friend, Martin Lister wrote as follows: "The way to find out the Nature of those *Magnetic Effluvia,* seems to be to enquire strictly into the nature of [them], and not to run giddily into Hypotheses, before we are well stocked with a Natural History—and a larger quantity of Experiments and Observations—which I think has hitherto been little heeded" (Lister, [1699] 1967).

7

David L. Hull

Scientific Bandwagon or Traveling Medicine Show?

I was originally asked by the editors of this volume to compare the reception of evolutionary theory in the nineteenth century with the reception of sociobiology today. Hindsight is a powerful tool in the hands of the historian. Because evolutionary theory turned out to be basically correct, anyone who opposed it in the nineteenth century must have been a closed-minded bigot. Hence anyone who opposes sociobiology today is equally obstructing scientific progress. In order to decrease the bias inherent in my investigations, I have added a third theory to the equation—an unsuccessful theory, phrenology. In the first half of the nineteenth century, Franz Josef Gall (1758–1828) suggested that there might be some correlation between the shape of a person's skull and his or her mental abilities. We now know that this view is nonsense. As might be ex-

This paper was prepared under National Science Foundation grant SOC 75 03535.

pected, phrenology degenerated into quackery. Hence its early supporters must have been gullible fools. But at the height of the movement phrenologists claimed among their numbers some of the greatest names of the day: John Quincy Adams, Prince Albert, Alexander Bain, Honoré de Balzac, Paul Broca, Charlotte Brontë, Henry Clay, Auguste Comte, George Eliot, David Ferrier, Karl Marx, Clemens Metternich, Edgar Allan Poe, and Mark Twain, not to mention four early evolutionists: Étienne Geoffroy Saint-Hilaire, Robert Chambers, Herbert Spencer, and Alfred Russel Wallace.

Both phrenology and evolutionary theory started off as genuine scientific theories. Serious scientists could be found arrayed on both sides of both issues. However, from our contemporary point of view, Gall lost and Darwin won. What did the phrenologists do wrong? What did the evolutionists do right? Could anyone at the time have been able to predict that evolutionary theory would succeed and phrenology fail? A strong tendency exists to conclude that Darwin's great achievement was to devise a theory that was basically correct and that Gall's failure was to come up with a set of ideas that were crudely mistaken. But what Gall *did* wrong was to *be* wrong. Scientific theories that contain a large element of truth continue to prosper, while those which are fundamentally in error either drop out of sight or degenerate into some form of pseudo-science. Hence, the message for the sociobiologists is "Be right!" If the views now being urged by the sociobiologists are reasonably close to the truth, the sociobiological bandwagon will turn into a victory parade. If not, it will degenerate into a traveling medicine show.

However, if the history of phrenology and evolutionary theory have anything to teach us, it is that the truth of new theories *as they are originally set out* is not all that important. Phrenology in the first half of the nineteenth century was no further from the truth than the theory of evolution, which became widely accepted in the second half. What really determines the success or failure of new scientific theories is how advocates of these views continue to conduct themselves. They must be conceptually flexible, socially cohesive, and terminologically rigid. The role of evidence in science is too obvious to belabor, but evidence never totally constrains the freedom of scientists in formulating their theories. The fudge factor is just as important (Westfall, 1973). Any scientist who is not a "master wriggler," to use Darwin's phrase, will see his views refuted almost immediately. Scientists can succeed only if they are willing to break a few methodological rules—sometimes every rule in the book. However, they cannot finagle at all costs. Falsifiability does matter in science but not the falsifiability of disembodied propositions. What really counts is the falsifiability of scientists. To be successful, a scientist must be able to recognize clear threats to his or her position and respond appropriately. But the proper response to imminent refutation is not admitting defeat; it is changing one's position while retaining one's original terminology. Successful scientists are those who master the art of judicious finagling.

Methodological Objections

The phenomenal increase in the number of learned journals has spawned a new literary genre—reviews of reviews. In a recent review of the reviews of E. O. Wilson's *Sociobiology: The New Synthesis* (1975a), Arthur Caplan (1976, p. 21) remarks that the critical responses "fall into two general categories. Some biologists have taken exception to Wilson's efforts for purely methodological reasons. Others, however, have objected to the book on moral or ethical grounds." (The accuracy of Caplan's conclusion can be confirmed by reading three recent review symposia devoted to sociobiology in the *American Journal of Sociology*, 1976, *82*, *Contemporary Sociology*, 1976, *5*, and the *American Psychologist*, 1976, *31*.) This same observation is equally true of both phrenology and evolutionary theory in the nineteenth century. Neither was judged to be properly scientific; both were deemed to pose a grave danger to humanity. In this section, I deal with the methodological objections raised to phrenology and evolutionary theory in the nineteenth century, comparing them to those currently being urged against sociobiology. In the next section, I do the same for the moral, ethical, and social objections. I conclude by seeing whether the differing fates of phrenology and evolutionary theory might not help us predict the future course of sociobiology.

According to the phrenologists, the brain is the organ of the mind. The brain can be analyzed into organs, the human mind into faculties, and an isomorphism can be established between the two. Furthermore, the relative size of each cerebral organ is a good measure of the power of the associated faculty. Finally, the external shape of the skull is an accurate reflection of the external shape of the brain beneath. Thus, by carefully examining a person's skull, one should be able to diagnose his or her mental capacities.

By now, the basic premises of Darwin's theory are commonplace. More organisms are produced than can possibly survive. A relation exists between the traits that an organism exhibits and the likelihood that it will survive long enough to reproduce itself. These traits vary from one generation to the next, and these variations in turn are heritable. Thus, evolution should take place, one species changing into another or splitting into two or more species. The one place where phrenology and evolutionary theory overlap is in their materialistic implications for mind. Minds cannot consist in separate, independent, immaterial substances.

In spite of the differences between these two theories, the methodological objections raised to them were the same. In the nineteenth century, "inductivism" was the official philosophy of science in Great Britain. In this view, true scientists begin by collecting facts wholesale without any preconceived notions. Gradually, low-level regularities emerge. After enough of these low-level regularities have been collected, more general regularities materialize out of them, and so on. Of course, none of the leading philosophers of the day actually held such views, even such empiricist philosophers as John Herschel (1792–1871) and John Stuart Mill (1806–1873), but everyone tried to sound as much as possible like Francis Bacon (1561–1626), while surreptitiously acknowledging roles in science for deduction, theorizing, and even speculation (Hull, 1973; Ruse, 1975a).

The methodological criticisms of phrenology voiced by P. M. Roget (1779–1869), the author of one of the Bridgewater Treatises (1834) and of the famed *Thesaurus* ([1852] 1965), are typical. Although Roget agreed that the brain is the organ of the mind and that several analogies lend an air of plausibility to the existence of some localization of brain function, he could go no further, objecting that "nothing like direct proof has been given that the presence of any particular part of the brain is essentially necessary to the carrying on of the operations of the mind" (Roget, 1842, p. 465). Analogies are fine for "directing and stimulating our inquiries to the discovery of truth by the legitimate road of observation and experiment. But to assume the existence of any such analogy as equivalent to a positive proof resulting from the evidence of direct observation is a gross violation of logic" (Roget, 1842, p. 466).

Numerous philosophers and scientists raised exactly these same objections to Darwin's theory. Adam Sedgwick (1785–1873) likened it to a "vast pyramid resting on its apex and that apex a mathematical point" (Sedgwick, 1860, p. 334). Darwin had deserted the "true method of induction," the "tram-road of all solid physical truth" (Darwin, 1899, 2, p. 43). As critical as Sedgwick was, Darwin still fared better than Robert Chambers (1802–1871), the author in 1844 of the popular *Vestiges of Creation*. Sedgwick (1845, p. 4) claimed that Chambers "has not so much as a mathematical point to rest his foot upon." The methodological objections to Darwin were set out in greatest detail by William Hopkins (1793–1866). While conceding that Darwin had the right to investigate the origin of species, Hopkins also gave Darwin to understand that "we exact from him in the support of his theories the same logical reasoning and the same kind of general evidence we demand before we yield our assent to more ordinary theories" (Hopkins, 1860, p. 739). But no one had ever directly observed the birth of a new species (nor has anyone to this day). Of all the experiments performed by animal breeders to produce varieties that were intersterile, "we are not aware of one which affords the slightest positive proof" (Hopkins, 1860, p. 80). Nor were Darwin's fellow Darwinians less critical. T. H. Huxley (1825–1895), for example, argued that the inductive foundations of evolutionary theory would remain insecure until the evolution of a new species was actually observed.

Needless to say, neither Gall nor Darwin presented much in the way of direct evidence or observational proof for their views. Their opponents objected that the connection between their "evidence" and their basic premises was extremely tenuous. They also complained of the modifications that Gall and Darwin made in their theories in the face of recalcitrant data. Gall originally postulated twenty-seven faculties, then raised the number to thirty-three. By the time his disciples had finished with the theory, they had postulated forty-four faculties and associated organs of the brain. In addition, one region of the brain could compensate somewhat for another. As the phrenologists added new faculties, dropped old ones, subdivided certain faculties, merged others, and remapped the regions of the brain accordingly, opponents such as G. Poulett Scrope (1797–1876) grew impatient, complaining, "Such is the kind of evidence on which is founded one of the most extraordinary theories that ever disgraced the unfortunate science of mental philosophy! By rapidly assuming the truth of his hypothesis, the phrenologist is capable of making a stand by means of that very complication and obscurity of his subject which ought to have been present in his mind at the first step of his progress. Once grant the existence of thirty-six organs, reciprocally acting on each other and influenced by adventitious circumstances, and he is a man of little ingenuity who cannot prove any possible arrangement of them to accord with the character of any given individual or provide a plausible account for the apparent discrepancy" (1836, p. 181).

Similar complaints were voiced on the "ductility" of the reasoning that Darwin used to defend his theory (for example, Hopkins, 1860). Such criticisms of both theories continue to this day, although the magic word is no longer *induction* but *falsifiability*. For example, Robert M. Young (1968, p. 254) concludes from studying the large compendium of evidence that Gall accumulated that "the phrenological method is a textbook case in support of a falsificationist view of scientific method, for he sought confirmations and failed to take exceptions seriously enough." Gertrude Himmelfarb (1959, p. 279) makes similar remarks in connection with Darwin's theory: "Thus when the actual evidence in the case proved to be insufficient for his purpose, Darwin referred back to the theory which he had found adequate in other cases—although it was precisely the adequacy of the theory that was being challenged." Darwin's argument "was too evasive and the reasoning too circular to satisfy all Darwin's champions, let alone his critics."

Throughout the history of science, scientists have claimed that they base their theories on the facts and nothing but the facts. Newton can be found proclaiming *"hypotheses non fingo"* ("I do not feign hypotheses"), the phrenologists claimed to deal only with "observed facts of nature," and Darwin (1899, *1*, p. 68) declared that he proceeded in developing his theory "on true Baconian principles and without any theory collected facts on a wholesale scale." Scientific mythology aside, none of these claims has the slightest foundation. What are the sources of this continued hypocrisy? One is that science as we know it developed in reaction to a research program that had been degenerating for some time. The dangers of building a system safely insulated from refutation by ad hoc hypotheses piled on ad hoc hypotheses were all too apparent to scientists struggling to get out from under scholasticism. Scientists wanted to emphasize the difference between their activities and those of scholastic philosophers. Such fears in the early years of science perhaps were justified. They have little justification today.

Another source for the hypocritical standards imposed on newly emerging scientific theories that is as relevant today as it was in Newton's time is the conception of theories as timeless sets of axioms, as if there were such a thing as *the* theory of evolution, *the* theory of phrenology, or *the* theory of sociobiology. If the theories that function in the ongoing process of science were such discrete axiomatic systems, then complaints about the ductility and circularity of scientific reasoning would be justified. Given a complete, explicitly formulated scientific theory, any particular observation statement will follow from it, or it will not. But actual scientific theories are much more variable and amorphous. It is not always clear which implications follow from a theory and which do not, because it is not always clear which tenets are central to the theory and which not. Scientists continually change their theories. Improvement of a theory in the face of a newly discovered difficulty can look very much like circular reasoning. In short, scientific theories evolve. Because they evolve the way they do, they have no essences.

From the beginning, evolutionists had to argue that species as units of evolution do not have essences. Species change indefinitely through time. At any one time, certain species might be characterized by sets of traits that are possessed by all the organisms that comprise that species and, collectively, *only* by these organisms. But such a distribution of traits need not be characteristic of all species, and if one follows species back through time even these gaps disappear. Species are not individuated on the basis of sets of essential traits, because typically these traits do not exist. More importantly, species are not even individuated on the basis of sets of statistically covarying traits. They are individuated on the basis of descent and evolutionary unity (Simpson, 1961; Ghiselin, 1974; Hull, 1976). Organisms that belong to the same species tend to be similar to each other, but they need not be and frequently are not.

The consequences of conceptualizing theories as historical entities are profound. For example, in the traditional view, there is some one set of axioms appropriately termed the "Darwinian theory of evolution." Richard Dawkins (1976, p. 210), in his highly readable book on sociobiology, expresses the essentialist position on the nature of theories with remarkable candor: "When we say that all biologists nowadays believe in Darwin's theory, we do not mean that every biologist has, graven in his brain, an identical copy of the exact words of Charles Darwin himself. Each individual has his own way of interpreting Darwin's ideas. He probably learned them not from Darwin's own writings, but from more recent authors. Much of what Darwin said is, in detail, wrong. Darwin, if he read this book, would scarcely recognize his own original theory in it, though I hope he would like the way I put it. Yet, in spite of all this, there is something, some essence of Darwinism, which is present in the head of every individual who understands the theory."

Nothing could be further from the truth. Darwin changed his mind on a variety of issues as he continued to develop his theory. He always maintained that species evolve gradually and that natural selection is the chief directive force, but he wavered on nearly every other issue. And no two Darwinians totally agreed with Darwin or with each other about the basic features of evolution. The evolutionary theory that became widely accepted in the nineteenth century was *not* any of the versions Darwin himself set out. The versions that were popular were typically progressive, saltative, and Lamarckian. If theories are interpreted essentially, then none of Darwin's strongest advocates were Darwinists: Huxley because he opted for saltative evolution; Asa Gray (1810–1888) because he preferred divinely directed progressive evolution. On the traditional view of theories, there was no Darwinian revolution. Darwin failed miserably, as miserably as Gall.

In response to the preceding claims about theories as historical entities, one might object that periodically scientists do claim that particular propositions are fundamental to their theories and, if falsified, would falsify their theory. For example, Darwin ([1859] 1964, p. 201) states that, if "any part of the structure of any one species had been formed for the exclusive good of another species, it would annihilate my theory, for such could not have been produced through natural selection." If taken at face value, statements such as these certainly make it appear as if theories are made up of essential premises and are easily falsifiable. In spite of Darwin's claim, a single instance of a structure in one species that serves the exclusive good of another would not annihilate "evolutionary theory." Even if it could somehow be shown that a structure contributed nothing to the organisms that possessed it, the legitimate conclusion is that natural selection is not the *sole* directive force in evolution—but Darwin repeatedly claimed that he never thought it was! Of greater importance, in the versions of evolutionary theory that actually became popular in the second half of the nineteenth century, natural selection played an even more subsidiary role.

Claims about "essential" tenets make theories look more static and falsifiable than they actually are. In reality, such claims serve to insulate theories against falsification. For example, William Hamilton (1788–1856) thought that he had decisively refuted phrenology by showing the frequency with which large sinuses exist between the cerebrum and the bones of the forehead, thus casting doubt on inferences from the shape of the forehead to the shape of the brain beneath (Cantor, 1975, p. 211). In response to objections such as these, Andrew Combe (1797–1847) argued that the only way to overthrow phrenology was to prove the negative of at least one of its basic tenets (Cantor, 1975, p. 213). But the propositions that are most basic to a theory also tend to be the most difficult to confront with evidence, a nice hedge against falsification. The fact that these tenets themselves can be secretly modified further insulates theories against easy refutation.

Exactly the same methodological objections that were raised against phrenology and evolutionary theory in the nineteenth century are currently being raised against sociobiology: "In order to make their case, determinists construct a selective picture of human history, ethnography, and social relations. They misuse the basic concepts and facts of genetics and evolutionary theory, asserting things to be true that are totally unknown, ignoring whole aspects of the evolutionary process, asserting that conclusions follow from premises when they do not. Finally, they invent ad hoc hypotheses to take care of the contradictions and carry on a form of 'scientific reasoning' that is untestable and leads to unfalsifiable hypotheses" (Allen and others, 1976, p. 182).

The methodological faults that the opponents of sociobiology find with it are real. Sociobiologists do reason in circles, occasionally taking for granted the very thing that they need to prove. Some of their hypotheses do look extremely suspicious. It is easy to appreciate the frustration of anyone trying to falsify the basic principles of sociobiology. What *are* these basic principles, and what conceivable evidence could possibly refute them? In the preceding pages, I have attempted to show that this state of affairs is common in science. The trouble with the methodological objections raised against sociobiology is that the standards from which they flow are applicable only to theories as finished products, and they are being used to assess a newly emerging theory.

Critics of sociobiology complain that no direct evidence has been presented for any of its basic premises. The notion of "evidence" for or against a view is problematic enough. For example, Marshall Sahlins (1976b) cites the very same class of social phenomena as counting against sociobiology that Alexander (1975, 1977) touts as its most persuasive confirmation—mother's-brother forms of kinship systems. The notion of "direct evidence" is even more problematic. In case after case in the history of science, the evidence demanded at the time that theory is introduced either is never forthcoming or else materializes long after the theory has been widely accepted. The appearance of clear, conclusive confirmations and refutations by unequivocal data is a function of our retrospective bias in favor of theories that succeeded (Holton, 1978). At the time, both Gall's and Darwin's interspecific comparisons were dismissed as not counting as evidence for anything. Today we agree that Gall's comparison of the heads of proud people and peacocks proves nothing. However, Darwin's comparison of the points that appear on the ears of some people and the pointed ears of other species implies, to some extent, common descent. Numerous nineteenth-century scientists called for the observation of one species giving rise to another. We are still waiting. We are also still waiting for the elimination of the gaps in the fossil record.

Sociobiologists have been accused of salvaging their position through the use of ad hoc hypotheses. Unfortunately, today's ad hoc hypothesis is tomorrow's law of nature. Numerous explanatory devices were dismissed as being ad hoc when they were first introduced: the *punctum aequans* of Ptolemaic astronomy (epicycles were all right), Heinrich Hertz's necessarily hidden bodies, Gall's complementation of cerebral organs, the epistatic genes of Mendelian genetics, and Wilson's "multiplier effect." However, there seems to be no correlation between how ad hoc an hypothesis seems when it is originally introduced and its eventual fate. Perhaps Wilson's multiplier effect will go the way of the *punctum aequans,* perhaps the way of epistatic genes. The same can be said for the "threshold effects" of the sociobiologists and the "genetic homeostasis" of their opponents. The survival of one theory over the other will determine which hypotheses were legitimate and which ad hoc.

One of the most telling criticisms of the sociobiological research program is the uncritical way in which sociobiologists extend to other species of organisms terms designed to refer to human beings and human societies. In ordinary discourse, we refer to Victoria and the fertile females in termite colonies as "queens," to Vietnam and conflicts between members of different ant hills as "wars," and so on. However, nothing about the scientific equivalence of these entities follows from ordinary usage. For example, the social structures that include human queens may have very little in common with termite societies and their queens. Queens in human societies on occasion have considerable authority; termite queens are hardly more than baby factories. (In this particular case, the contrast between termite queens and Victoria is not as great as one might wish.) Hence, the critics conclude that sociobiology rests on "mere analogies" and "mushy metaphors."

Once again, this situation is common in science. Phrenologists adopted commonsense human faculties such as combativeness, covetousness, and cautiousness and then analyzed the brain into organs accordingly. But in this respect they are no different from early Mendelian geneticists. They too adopted commonsense divisions of organisms into traits—blue eyes, green seed coat, cleft chin, and so on—and then analyzed the hereditary material into genes accordingly. Phrenologists used the shape of skulls to infer the shape of the brain beneath. Only after death could they inspect the brain itself. Mendelians were in even tighter straits. They had nothing like direct access to genes. Instead, they had to postulate genes as a result of following patterns of inheritance for phenotypic traits. The practice of correlating commonsense, observable traits with theoretical, inferred entities is fundamental to theoretical science. The difference between the phrenologists and the Mendelians is the alacrity with which the Mendelians were willing to abandon their original crude classification of traits into kinds. For example, elliptocytosis in human beings was once considered *a* trait. When two different genes, each capable of producing this effect, were discovered, it was promptly divided into two different traits. A few phrenologists, such as David Ferrier (1843–1928), followed a similar pattern, sacrificing functions to physiological accuracy, but most, such as Gall, allowed his catalogue of functions to dictate the organization of the brain (Young, 1968, p. 260).

The fate of sociobiology will depend in part on how adept sociobiologists are in redefining their commonsense classifications, turning them into technical scientific classifications. The term *work* in ordinary English has very little to do with *work* in physics. Before the sociobiologists are done, *aggression, altruism,* and *dominance* in ordinary English may have little in common with these terms as they are used in sociobiological theory. The process is already well underway. Sociobiology does lean heavily on various analogies. So did Darwin's theory (Young, 1971; Ruse, 1973, 1975b; S. J. Gould, 1976a). Whether these analogies are dismissed as mere analogies, retained as useful, or promoted to literal truths will depend on the success of the sociobiological research program.

Ethical Objections

At first, one might be surprised to discover a scientific theory being condemned for its moral, ethical, and social implications, but the practice has been common throughout the history of science. In each case, the villains have been mechanistic, materialistic, atomistic, and reductionistic theories, and their sin has been the destruction of the image that human beings had forged of themselves. If we do not live in the center of the universe, if we evolved instead of being specially created, if causes operate in the mental world as well as in the physical, if our actions are totally a function of our environment and our genes—then we are nothing. In discussing these charges, one fact frequently is overlooked: By and large, they are justified. Thus far, the most successful scientific theories have been mechanistic, materialistic, and so on, and the acceptance of these theories has destroyed the image that human beings have had of themselves. In the nineteenth century, evolutionary theory destroyed the faith of millions. It still poses a serious threat to the faith of millions more. Theories that concern the connection between our brains, minds, and behavior have been equally threatening. The cries currently being raised about the doctrines of sociobiology may also be justified. The dangers that science has posed to various systems of belief throughout history have proven time and again to be very real.

However, before the dangers that sociobiology poses to humankind can be discussed, three distinctions must be made: One is *rejection* versus *suppression,* another is the *truth* of a claim in contrast to how *dangerous* it might be, and yet another is the difference between the implications that a scientific theory has for *moral principles* and for *social policy.* The first distinction turns on how publicly and openly the view in question is discussed. In order to reject a view, regardless of the reasons for rejecting it, one must be aware of it. The notion of truth is extremely problematic, but at least it is reasonably distinct from the notion of danger. Both true claims and false claims can be dangerous if they are believed and acted on. However, if one is a realist, the presumption is that believing something that is false is always dangerous. Being mistaken about the world in which we live is always liable to lead to trouble. Finally, if one accepts the traditional philosophical view that "ought" cannot be derived from "is," then a scientific theory might set constraints on social policy, but it cannot determine which acts are moral and which immoral.

Scientists frequently claim that an hypothesis should be rejected because it is false. When the reasons for rejection are scientific, as I have argued in the previous section, decisions are difficult enough to make. When the reasons given for rejecting a scientific theory are extrascientific, scientists are placed in an even deeper quandary. Can extrascientific considerations legitimately enter into a scientist's decision to accept or reject a scientific theory? Should scientists reject a theory because it is incompatible with the formal logic of its day, because it is too complicated, or because it is ugly or evil? The issue is the proper domain of science and its insulation from the rest of human endeavors.

From the beginning, scientists have attempted to protect themselves by a variety of devices. If their views conflicted with current orthodoxies of one kind or another, they could always claim that they were not trying to explain how the world actually is. Instead, all they have been trying to do is to produce calculation devices that allow for more accurate predictions. In other ages, the insulation was accomplished by asserting that scientific theories properly interpreted could not possibly conflict with Scripture properly interpreted. Or one might claim that natural phenomena as they are taking place now are the proper domain for science, but not the mysteries of creation. Or one might argue that science deals with the physical world, not the living world, or the thinking world, or the world in which souls reside. More recently, the positivists have argued for the exclusion of metaphysics from science. Finally, as contemporary wisdom has it, morals at least are totally separate from matters of fact.

In each case, the protective barricade was abandoned once it had served its purpose. Scientists currently are realists: Scientific theories are supposed to tell us what the world in which we live is like. Any good book that foolishly makes factual claims about the empirical world runs the risk of being contravened by scientific investigations. Scientists now deal as freely with questions of creation as with dynamical issues. The domain of science has been steadily enlarged to include biological, psychological, and social phenomena. Science is even beginning to threaten the few remaining sanctuaries of human endeavor in the form of evolutionary epistemologies, naturalistic ethics, and the like. The defenders of the absolute diremption between science and morality have a right to feel uneasy. Philosophers and, to a lesser extent, scientists now acknowledge the inextricability of science and metaphysics. Perhaps morals are next. Perhaps we will yet see the day when philosophers and scientists alike maintain that a scientific theory should be rejected as false because it conflicts with some moral tenet or that some moral tenet should be rejected as evil because it conflicts with a currently accepted scientific theory.

That day, however, has not yet come. Even though one version of Marxism (and Marxism comes in as many versions as Christianity and the synthetic theory of evolution) grounds morality in objective reality, all of the participants in the debate over sociobiology seem to agree that no moral implications can be drawn from a genuine scientific theory, and vice versa. However, certain critics insist that scientific theories can have implications for social policy—implications about how society *can* be, rather than how it *should* be. Perhaps everyone in a society should be educated, but if people with twenty-one trisomy (Mongolian idiots, as they used to be called) are not educable, then there is no point in wasting the resources of a society attempting to educate them. When the issue is moral implications, scientists are protected by the is-ought distinction. When the issue is the application of scientific knowledge, scientists seek the protection of the theory-practice distinction. Their task is to seek knowledge for its own sake. Knowledge, of course, can be used for both good and evil. Either way, theoretical scientists are not responsible. In point of fact, however, this is not how scientists have conducted themselves in the past. They have been happy to take the responsibility for the *good* applications of scientific knowledge, accepting Nobel Prizes, knighthoods, and the like. Consistency would seem to require that they also accept some of the responsibility for the bad applications.

In growing numbers, scientists are beginning to acknowledge that they carry more responsibility for the applications of scientific knowledge than do ordinary citizens, but ample disagreement exists about the extent of this responsibility. Should a scientific theory be suppressed because it is dangerous, even though it might be true?—that is the question. The slightest hint of suppression is sure to elicit howls of outrage from the scientific community, and understandably so. From the beginning, scientists have suffered at the hands of those who would determine which scientific views could be made public on extrascientific grounds. Most of us are likely to find such suppression villainous. At the beginning of the nineteenth century, the ecclesiastical authorities brought sufficient pressure to bear on the Austrian government to force it to prohibit Gall from presenting his views on phrenology to the local citizens, an action that seems only to have increased the popularity of phrenology. At roughly the same time, William Lawrence was forced to withdraw from circulation two books which he had published because of their materialistic implications for the human species (Wells, 1971). However, by the second half of the nineteenth century, the most that William Whewell (1794–1866) could do was to ban Darwin's *On the Origin of Species* ([1859] 1964) from the shelves of Trinity College, where he was Master.

Phrenologists and evolutionists alike used attempts to suppress their views to their own advantage. Like Galileo, they were victims of religious persecution. Even Roget, one of phrenology's sternest critics, condemned the "senseless clamor" against phrenology by "bigoted priests" (1842, p. 456). J. D. Hooker (1817–1911), looking back at the early reception of evolutionary theory, was no less bitter: "It is not for us, who repeat *ad nauseam* our contempt for the persecutors of Galileo and the sneers at Franklin, to conceal the fact that our great discoverers met the same fate at the hands of the highest in the land of history and science" (Huxley, 1918, *2*, pp. 302–303).

In 1864, a group of young scientists circulated a declaration among their fellow scientists, asserting that science had no business contradicting revealed religion. Science and Scripture, once properly construed, could not possibly conflict. They were able to gather only 717 signatures; of these only a handful were those of especially well-known scientists. Their biggest catches were David Brewster, James Prescott Joule, and Adam Sedgwick. The response of C. G. Daubeny (1795–1867) was typical of the scientists who refused to sign the declaration: "I am not prepared to declare that conclusions honestly arrived at in the course of scientific investigations, as for instance those relating to the age of the world, the antiquity of the human race or the prevalence of a deluge over parts of the globe not at the time inhabited by man ought, as a matter of Christian duty, to be suppressed and ignored" (Daubeny, 1864, p. 9).

Current attempts to suppress sociobiology because of the dangers it poses to humanity have met with the same response: "The central issue is whether scientists should have the freedom to explore nature no matter how badly their discoveries may be misused or how sharply they may contradict contemporary orthodoxies. It is an argument at least as old as Galileo's clash with the church, and the evidence of history is that the forces of suppression do not prevail for long" (Panati, 1976, pp. 51–52).

However, scientists themselves frequently act in ways that effectively suppress ideas with which they disagree. A conspiracy of silence is the scientific community's most effective weapon. In the case of both phrenology and evolutionary theory, the initial response of the scientific community was silence. The more prestigious a society and its journal, the more impenetrable the silence. As Frederick Burkhardt (1974, p. 33) notes, "Between 1859 and 1870, no paper in either the *Proceedings* or the *Philosophical Transactions* of the Royal Society has as its subject a discussion of Darwin's theory of the origin of species." From Gall and Darwin to Velikovsky and Wegener, a common response of the scientific community to ideas that they take to be hopelessly mistaken is an attempt to suppress them by ignoring them. Anyone may publish anything he or she pleases, but not in a scientific journal, and scientists are under no obligation to take note of ideas if they choose not to. The end result can be fairly effective suppression. For champions of orthodox views, the best response to an attack is initially no response at all. Only if the attacks begin to attract converts can powerful scientists be smoked out to defend themselves. Not too surprisingly, E. O. Wilson's colleagues advised him that the "best response to a political attack . . . is perhaps no response at all" (Wilson, 1976a, p. 183). No doubt his critics were placed in the same position. Any attack, but especially a political attack, serves only to bring further attention to views that they do not want to see disseminated.

Some of the critics of sociobiology seem to argue for the *permanent* suppression of its doctrines *regardless* of their truth. However, most take a more moderate view. They argue instead for raising the standards of proof when the ideas addressed are socially explosive. For example, S. J. Gould (1976a, p. 22) states, "I make no attribution of motive in Wilson's or anyone else's case. Neither do I reject determinism because I dislike its political usage. Scientific truth, as we understand it, must be our primary criterion. We live with several unpleasant biological truths, death being the most undeniable and ineluctable. If genetic determinism is true, we will learn to live with it as well. But I reiterate my statement that no evidence exists to support it, that the crude versions of past centuries have been conclusively disproved, and that its continued popularity is a function of social prejudice among those who benefit most from the status quo."

The same observations could have been made against Darwin's work, and they were. In response to Darwin's *The Descent of Man* ([1871] 1969), one writer in the *London Times* (April 8, 1871, p. 5) proclaims, "A man incurs grave responsibility who, with the authority of a well-earned reputation, advances at such a time the disintegrating speculations of this book. He ought to be capable of supporting them by the most conclusive evidence of facts. To put them forward on such incomplete evidence, such cursory investigations, such hypothetical arguments as we have exposed, is more than unscientific—it is reckless."

Thus the two classes of objections to sociobiology with which this chapter deals—methodological and ethical—converge on each other. The opponents of sociobiology apply uncommonly high standards of proof to it, but not because they have unreal views about scientific proof. The higher standards result from the reputed danger of certain sociobiological doctrines. If one admits that scientists have some responsibility for the use and abuse of their ideas, the preceding position has much to recommend it. It is certainly rational to raise one's standards of proof when the probable outcome is serious. Rational people demand more evidence of malignancy before removing a kidney than a mole.

However, this position itself is beset with difficulties. In the first place, when the claim that socially dangerous scientific hypotheses should not be made public is coupled with the prohibition of research on socially dangerous hypotheses, the net effect is the total prohibition of "dangerous ideas." Secondly, there is the problem of deciding which ideas are dangerous. Scientists have proven themselves to be extremely good at making decisions regarding empirical fact. But there is no reason to suppose that they are any better than the rest of us in making moral decisions, and the track record of humanity at large on this score is not impressive. The critics of sociobiology assume that the ideal society would be egalitarian, cooperative, and socialistic. But others see nothing wrong with elitism, individualism, and capitalism. To the rugged individualist, the reputed individualistic implications of sociobiology are not in the least alarming.

The critics also assume that there is some connection between the actual content of a theory and its social implications, both licit and illicit. Phrenologists should have been "hereditarians." They tended to be "moderate environmentalists." In Edinburgh, phrenologists tended to champion liberal causes; so did their opponents. On the continent, phrenologists were more conservative (Young, 1968; Parssinen, 1974; Cantor, 1975; Shapin, 1975; Cooter, 1976; Cowan, 1977). By and large, Europeans were racist before the publication of *On the Origin of Species*. They remained racist afterward. It is no more difficult to justify the inherent superiority of Caucasians by reference to evolutionary theory than it is by reference to Scripture (Haller, 1971). In connection with phrenology, Steven Shapin (1975, p. 241) concludes, "Pure hereditarian ideas may serve as well as pure environmentalist ideas in legitimating tolerance, liberality, and social justice." Similarly, Noam Chomsky (1975, p. 132) points out that extreme environmentalism can be abused as readily as extreme hereditarianism. And, to round out the story, B. F. Skinner (1974) has repeatedly argued that the prevalent belief in free will has also contributed its share to human misery.

To anyone committed to the efficacy, however slight, of reason, argument, and evidence in the course of human affairs, the ease with which scientific theories can be bent to support almost any social cause whatsoever is discouraging, but the problem is not that the implications are social. The same difficulties arise in deciding which observational consequences follow from scientific theories and have the same source—the evolutionary character of scientific theories. Is sociobiology racist and sexist? It is difficult to say. Would the presence in a society of a high percentage of children being raised by people not related to them genetically refute sociobiology? It is just as difficult to say. Such quandaries are common in science. Were phrenology and evolutionary theory racist and sexist? Did the gaps between the cerebrum and the skull refute phrenology any more conclusively than the even larger gaps in the fossil record refuted evolutionary theory? The distinction between licit and illicit implications of scientific theories is far from sharp.

Finally, those who advocate increased standards of support for scientifically dangerous ideas before they are made public are themselves not immune from this same line of reasoning. The suppression of scientific ideas itself carries considerable risk. In the past, all the wrong people seem to have attempted to suppress new scientific ideas for all the wrong reasons. Those who would suppress scientific ideas for moral reasons today are under some obligation to set out the moral standards on which these decisions are based and to justify them. This particular dispute is too important for it to continue on the basis of tacitly held moral principles.

Conclusion

I began this chapter with the promise that studying the fates of phrenology and evolutionary theory in the nineteenth century might help us assess the current status of sociobiology and predict its future. It is now time to deliver on that promise. The same methodological objections were raised to phrenology and evolutionary theory. Both were condemned as being subversive to humanity. Yet phrenology lost, and evolutionary theory won. The same methodological objections are now being raised against sociobiology, and it too is being condemned for being inhumane. Are either of these sorts of objections likely to make a difference to its success or failure? T. M. Parssinen (1974) has identified four reasons for the decline of phrenology in nineteenth-century Edinburgh: (1) its popularity with the working classes, (2) the crumbling of its scientific basis because of the experimental investigations it encouraged, (3) the disintegration of its organizational structure over such issues as materialism, and (4) the drifting of its members into the study of mesmerism and phrenomesmerism.

Parssinen's list of reasons for the demise of phrenology is interesting in two respects. First, no mention is made of methodological or humanitarian shortcomings. As large as these objections loomed in the critical literature, Parssinen thinks they had little effect. In the first section of this chapter, I tried to explain why methodological criticisms have been so ineffectual in science. One reason is obvious: At any one time in the history of science, scientists have a set of stock objections that they raise to any new theory, regardless of its content or merits. These stock objections tend to be both extremely empirical and highly hypocritical. They are applicable, if at all, only to scientific theories as finished products. They are totally useless in assessing newly emerging theories.

A more subtle reason for the ineffectiveness of methodological criticisms is that the merits of a particular scientific method are determined in large measure by its success. There is nothing especially "unscientific" about introspection. Its poor reputation results from the fact that its results are highly variable, undependable, and idiosyncratic. Scientists have used all sorts of methods in their investigations: observation, intuition, divination, revelation, analogy, common sense, and so on. Which of these methods are genuinely scientific cannot be determined *a priori*. If the "wow!" feeling of discovery were more dependable, confirmation might be much less important. Or, if divine revelation guaranteed truth, scientists might still be using it. Of course, in such circumstances, the "science" that would have resulted would not be science as we know it. As I see it, philosophy of science is an empirical, not a linguistic, undertaking. Our task is not to analyze the notion of science as it is currently understood—assuming that such a notion exists—but to reform our understanding of science as we come to learn more about it. Any philosophy of science that makes the great achievements in the history of science ill founded, illogical, and irrational must be seriously in error. Scientists are far from infallible, but they are doing something right. In short, I am arguing for a retrospective legitimization of scientific reasoning on the basis of the success of that reasoning. To the extent that sociobiologists are successful, their modes of reasoning will become part of legitimate scientific method. To use Dudley Shapere's felicitous dictum, "In science, we learn *how* to learn *as* we learn."

The influence of the moral character of scientific theories on their reception is more difficult to assess. Extrascientific considerations have had noticeable effects on the success or failure of scientific ideas. One of evolutionary theory's strongest selling points was that it was naturalistic. No miracles were needed. It is also no accident that the form of evolutionary theory that became popular was progressive. Progressivism was pandemic at the time. How much of a role did the hereditarian implications of phrenology and evolutionary theory play in their receptions? In the second section of this chapter, I have tried to explain why such questions are so difficult to answer. In contrast to the iconic representations assumed by traditional philosophies of science, actual scientific theories are extremely amorphous and variable. At any one time, numerous different versions of the same theory coexist, and these clusters change through time. Although the connection between scientific theories and their environments is not chaos, it is extremely complicated and variable. Given the current state of the art, we are in no position to state with any justification the relative importance of various "external" factors in the reception of newly emerging scientific theories. The only study to date that is sufficiently extensive to provide some basis for judgment can be found in Thomas Glick's *The Comparative Reception of Darwinism* (1974), and the only conclusion that it supports is that such connections are extremely variable.

A second feature of Parssinen's list of reasons for the demise of phrenology in nineteenth-century Edinburgh is that only one—the crumbling of its scientific basis—is the sort of reason that traditional philosophies of science consider legitimate. The others concern the sociology of science. If they are operative, they should not be. Reason, argument, and evidence are supposed to be the determining factors for scientists opting one way or the other on scientific issues. So far no one has presented much in the way of evidence that scientists actually *do* behave the way they *should* behave, but, more importantly, evidential and sociological considerations are not as distinct as philosophers have traditionally supposed.

Parssinen mentions the popularity of phrenology among the operative classes, the implication being that popularity among the wrong sorts of people can hurt the reputation of a scientific theory. It is certainly true that phrenology was popular among the working classes and that phrenologists were anxious to convert the operative classes, as indicated by Johann Caspar Spurzheim's (1776–1832) lectures to the London Mechanics' Institution (Parssinen, 1974). But Darwinism was just as popular among the lower classes. Huxley's lectures to working men were a huge success. The popular press is currently paying considerable attention to sociobiology, and the sociobiologists themselves have gone so far as to prepare a high school textbook (De Vore, Trivers, and Goethak, 1973). But if the fates of phrenology and evolutionary theory imply anything on this score, it is that public popularity is neither necessary nor sufficient to guarantee acceptance by the scientific community. It is not lethal, either. Conducting scientific debates under the public eye is more difficult; it is not impossible.

The effects of the other three factors that Parssinen lists as being relative to the failure of phrenology are clearer. The Darwinians were able to tolerate considerable doctrinal divergence while maintaining sufficient social cohesion. The phrenologists were not. Even though Wallace came to advocate phrenology, spiritualism, and a form of radical socialism and withdrew from extending the principles of evolutionary theory to human beings, he always remained a Darwinian and on good terms with Darwin. Huxley differed strongly over the prevalence of saltative evolution, yet remained firmly in the Darwinian camp. However, at a time when phrenology had few active and influential converts, Spurzheim broke with his master, commenting that thereafter Gall ceased to do any original work. The raising of the materialistic implications of phrenology led to wholesale defections (Parssinen, 1974, p. 13). Although the Darwinians were just as divided over such issues as materialism, they did not allow these disagreements to destroy the cohesiveness of the movement.

Externalists have claimed that the Darwinian clique won because its members were extremely well placed when the dispute over evolution broke out and remained so. For example, in the early years after the publication of *On the Origin of Species,* the presidency of the British Association for the Advancement of Science (BASS) alternated between Darwinians and anti-Darwinians (Ellegård, 1958). But several early phrenologists were also reasonably well placed in the scientific commmunity—for example, Étienne Geoffroy Saint-Hilaire (1772–1844) and George S. Mackenzie (1780–1848)—and several others were to become so—Alexander Bain (1818–1903), Herbert Spencer (1820–1903), Alfred Russel Wallace (1823–1913), and Paul Broca (1824–1861). However, phrenology as a scientific movement was unable to integrate itself into the larger power structure of science. For example, in 1834 when the British Association for the Advancement of Science turned down the request of George Combe (1788–1858) for a section on phrenology, Combe was forced to organize the Phrenology Association, which met at the same time and in the same city as the BAAS.

Internalists, on the other hand, emphasize the contributions that the evolutionists made to the substantive content of science. Many Darwinians had produced scientific work of recognized excellence before their conversion and continued to do so afterward, but from an "evolutionary perspective." Several early phrenologists also made substantive contributions to science, not the least of which were those of Saint-Hilaire, Bain, Spencer, Wallace, Broca, and Ferrier. Many of the scientific contributions of phrenologists did not bear directly on the principles of phrenology, but the same can be said for the evolutionists. Advances *were* made in the early years of the nineteenth century in understanding the brain, but they were made by the enemies of phrenology, not by the phrenologists. For example, in 1829 Sir Charles Bell (1774–1842) won the first Royal Medal for his work on the nervous system. Much later, Broca and Ferrier presented evidence for cerebral localization in the name of "scientific phrenology" (Young, 1968), but apparently they were too late. Phrenology's reputation had been fixed. In the ninth edition of the *Encyclopaedia Britannica,* Macalister (1888, p. 847) concedes that the "doctrine that the brain is the organ of the mind is now universally received" and that there "is a large weight of evidence, which cannot be explained away, in favor of the existence of some form of localization of function." But he is able to assert that "no phrenologist" contributed any original information on these points.

The question thus becomes "Why, in retrospect, do we view phrenology as having failed and evolutionary theory as having succeeded?" Both contained basic premises that we now take to be true. Both were modified extensively as they developed. Both attracted adherents, among whom were well-placed scientists, some of whom continued to contribute to the substantive content of science in the name of their respective research programs. Part of the answer is differences in number. For every successful phrenologist, there were hundreds of evolutionists. Another turns on scientific theories and research communities being historical entities developing through time. Like species, they are individuated on the basis of descent and unity, not similarity. If asked why two versions of a scientific theory are versions of the same theory, we are likely to answer in terms of extensive overlap in the claims that these two versions make. However, neither historians of science nor scientists themselves individuate theories in this way. Today a growing number of evolutionists (for example, Eldredge and Gould, 1972) are adopting a view of evolution that in several important respects is much more like the version set out by Richard Goldschmidt (1878–1958) than like that of Darwin, yet they consider themselves neo-Darwinians, not neo-Goldschmidtians. Why? Because they trace their views back through such neo-Darwinians as H. L. Carson and Ernst Mayr to Darwin, not to Richard Goldschmidt.

As a result, two formulations can be extremely similar with respect to substantive assertions and not count as versions of the same theory, while two formulations can be substantively very different and still be considered versions of the same theory. For example, in Darwin's day Huxley and Gray were considered Darwinians, and their versions of evolutionary theory were considered merely modifications of Darwin's views. Saint George Jackson Mivart (1827–1900) set out a non-Darwinian theory of evolution that did not differ materially from the views of Huxley and Gray. It was saltative, like Huxley's, and progressive, like Gray's. The substantive content of science is, of course, important. Scientists are more explicitly and consciously concerned with it than with anything else. But it does not follow on this account that scientific theories should be individuated in terms of similarity in substantive content (Hull, 1975, in press; Burian, 1977).

If future historians and scientists are to look back on sociobiology and conclude that it succeeded, then sociobiologists had best busy themselves on several fronts. They should stay active in science, both sociologically and substantively. The actual connections between the contributions that they make and sociobiology are not all that important as long as these contributions are recognized as coming out of the sociobiological research program. It would also help if they—not their opponents—made advances in understanding the biological bases of social behavior. They must be willing to modify their views, no matter how extensively, in the face of various considerations that threaten to undermine their position, while maintaining that these are *modifications,* not *abandonments.* The results are new *versions* of their theory, not *replacements.*

That selection occurs only at the level of the "individual" is supposedly one of the basic premises of sociobiology. If all adaptations can be explained solely by reference to individual genes, then the sociobiologists are right. If recourse to organisms must be made, then claims about selection operating exclusively at the level of genes are merely idiosyncrasies of particular sociobiologists. If kinship groups can function as units of selection, then kinship groups are individuals. If populations can function as units of selection, that is individual selection too. Populations and possibly entire species form individuals (Ghiselin, 1974; Hull, 1976). If this fails, then include group selection among the tenets of sociobiology.

Looking back, all the versions of evolutionary theory set out in the nineteenth century were seriously flawed. Darwin thought that evolution was gradual, undirected, and guided primarily by natural selection. Today we agree that it is undirected and that natural selection is the primary mechanism for evolutionary change.

However, we disagree with Darwin over the subsidiary mechanisms, and a growing number of evolutionists are opting for nongradualistic forms of evolution. Other versions of nineteenth-century evolutionary theory fare no better when compared to contemporary versions. The basic premises of phrenology fare no worse. The chief failing of the phrenologists is that they were not sufficiently adept at finagling. Instead of adapting the principles of associationist psychology to those of phrenology, Bain and Spencer defected. Evolutionists were masters at transmuting refutations into confirmations. Phrenologists allowed themselves to be saddled with the weakest part of their program (the reading of heads) instead of the strongest parts (for example, brain localization). Of course, the reading of heads was also the most original part, but comparable observations can be made with respect to evolutionary theory. The weakest and most original part of Darwin's theory was natural selection. That did not keep Darwin's disciples from saving their program at the expense of Darwin's unique contribution. The only form of evolutionary theory that the scientific community was willing to accept was progressive; evolutionists promptly produced a progressive version of evolutionary theory and labeled it "Darwinian." If current versions of sociobiology sound too inhumane, they can always be made to seem more attractive. There is no reason why a book must be named *The Selfish Gene* (Dawkins, 1976). *The Altruistic Organism* would do as well and be no more misleading.

In the long run, the success of sociobiology will depend more on how adaptable the sociobiologists turn out to be and how tenaciously they fight for the term *sociobiology* than on how close their current views are to the truth. One frequently hears rumblings of discontent over the term *sociobiology* and the attention paid to Wilson, but the sociobiologists are stuck with the term and their titular leader. Abandonment of the term and attacks on the man will register as a repudiation of the entire enterprise. The animosities and disagreements that arise among the members of a scientific movement are frequently far more bitter than those that arise between competing groups. If a research program is to succeed, these disputes had better stay intramural. The social cohesion of a research community contributes as much to the success of a research program as does conceptual unity, especially in times of rapid and fundamental conceptual change. In the end, future so-ciobiologists may agree with none of the claims made by the found-ers of the movement. In fact, their position may differ in no im-portant respect from that currently held by its opponents. If they call themselves *sociobiologists* and couch their position in the ter-minology of sociobiology, however, sociobiology will have won (for further discussion, see Hull, 1978).

8

John R. Searle

Sociobiology and the Explanation of Behavior

*A*s I am about to make a few negative remarks about sociobiology, let me begin with an expression of enthusiasm for its pros-
pects. I find its synthesizing aspects—that is, its ability to assimilate
a great deal of diverse data about different organisms under a single Darwinian explanatory model—very exciting. And I do not
agree, incidentally, with those who claim that it encounters substantial opposition because people find it revolting or repulsive to
be told that we are continuous with the rest of animal and plant
life. On the contrary, it appears to be a feature of the *Zeitgeist* that
many of us find that idea very attractive, and it is an ironic feature
of this particular intellectual revolution that, while its proponents
keep telling us how much opposition they are facing from all those
people who dislike being told they are animals, they in fact exploit
the current fashion of treating *Homo sapiens* as just another animal.
Many of us find it appealing to be told we are animals, and we find

it easy to empathize with all those pictures in the textbooks of animals fighting, making love, and looking after their infants, because we find them, in a word, all too human.

　　As intellectual revolutions go, however, sociobiology is a bit peculiar, because it is hard to find any really original ideas in it. I find two basic axioms or assumptions behind the sociobiological work that I have read, and I want to discuss those two. The first is that the phenotypes of human and animal social behavior are, at least to some extent, determined by the mechanisms of natural selection. This is an extension of Darwin's conception of natural selection; but is it not, after all, an extension that has been made before? There are some original wrinkles put on this idea in the discussion of such matters as the operation of kin selection in the evolution of altruistic behavior, but the underlying idea is not new. The second assumption (and I get the impression that many sociobiologists see it as a natural extension of the first) is that all animal behavior, and therefore all human behavior, is mechanically determined by genetic and environmental factors. By "mechanically," I mean that there is no intrinsic mental element in the causal explanation of the behavior. This assumption also is not new; in fact, it is as old as philosophical materialism and determinism. These two assumptions together give sociobiology both its objective and its explanatory paradigm within that objective. According to the first, the main aim of sociobiological investigation is the application of Darwinian natural selection to human and animal behavior. The working hypothesis is that wherever behavior has any element of genotype it is likely to be the result of natural selection. According to the second, in explanations of human and animal behavior there should be no reference to purposes, needs, and goals or to mental states, such as beliefs, fears, hopes, and desires. In this discussion, I want to propose that sociobiologists can keep the first of these assumptions while abandoning the second and that some of them mistakenly take the first as necessarily excluding mentalistic explanations.

　　It is part of the achievement of biology in general and of evolutionary biology in particular that they have been able to account for a variety of apparently teleological or goal-directed phenomena in terms that are purely mechanistic. Indeed, what is so

marvelously appealing about Darwinian theory is that it reduces
the appearance of purpose and teleology to mechanical explanation. Take a simple example. It is tempting to suppose that plants
turn in the direction of the sun because they need sunlight in order
to perform photosynthesis, on which their survival depends. However, for the appearance of teleology in such a case biologists are
able to substitute mechanism. They point out that the actual cause
of the plant's movement is the secretion of auxin, and they further
point out that plants that secrete auxin in this way will get more
sunlight and therefore have a greater chance of survival than similar plants that do not. Their ability to secrete auxin is, in a word,
adaptive; and adaptive traits survive. Biological species look as if
they had been deliberately designed for certain purposes, and indeed they seem to function as if they had certain purposes in mind,
with survival being the main purpose of all. But this is just an appearance, and it plays no role whatever in the actual causal explanation of the phenomena. The actual explanation is a combination
of (1) the causal role of the phenotype in the survival and in the
reproductive capacity of the organism and (2) the evolution of the
underlying genotypes by means of natural selection. It is perfectly
all right for us to talk about the plant as "needing" sunlight and
"seeking" the sun, but it is crucial to remember that such talk is
just a manner of speaking and that reference to purpose plays no
role in the explanatory apparatus of biology.

When we turn from plant to animal behavior, we find that
the explanatory mechanisms used are exactly the same. For example, consider the question "Why do certain species of birds migrate from the colder part of the United States in the winter to
warmer southern climates?" "Well," you might say, "the underlying
reason is that the cold weather is miserable and unpleasant for the
birds; it is hard to stay alive in such temperatures, so they go south
where the living is easier." But sociobiologists employ the same
explanatory apparatus here that plant biologists employed in the
plant case. They point out that the migration is adaptive. Animals
that make the migration have a greater chance of survival than
those that do not. The birds do not go south in order to stay alive;
rather, they have stayed alive because they are the sorts of birds

that are going to go south in the winter anyway. And they are the sorts of birds that go south in the winter because they have certain mechanisms, such as light-sensitive hormonal balances in portions of the hypothalamus that determine them to migrate, and these are inherited traits. In this account of the migration of birds and the light-seeking behavior of plants, all reference either to purpose (teleology) or mental states (intentionality) is eliminable. All behaviors are phenotypes, and all phenotypes are completely determined by a combination of the genotypes and the environment.

Actually, if you look closely you will see that there are two elements in this explanatory model corresponding to the two assumptions I mentioned earlier. The first element is the Darwinian hypothesis that those phenotypes that are adaptive and that have a genotypic basis survive by natural selection. The second is the postulation of mechanism: All behaviors, whether the turning of a plant or the migration of birds, are determined by physical mechanisms. I believe that it is in part because the Darwinian revolution was successful in showing that the appearance of mental intervention in the development of species could be completely accounted for on mechanical hypotheses that some sociobiologists were erroneously led to the conclusion that mental states play no causal role in the explanation of the behavior of specific animals within the various species. Whether or not I am right in this diagnosis of the origin of their beliefs, some of them at least do oppose mentalism in explanation as if it were inconsistent with their research program and possibly even unscientific. The position is stated very succinctly by David Barash in the following passage, where the two assumptions are run together: "It should be emphasized here and remembered throughout all evolutionary analyses of behavior that no assumptions need be made about the internal motivational state of the individuals concerned. Thus it is a convenient shorthand to use such expressions as 'concerned with,' 'has an interest in,' 'is better off by doing,' or even 'wants to.' These expressions do *not* imply cognition or volition. They are simply less clumsy than saying, 'has been selected for responding in such a manner because ancestors behaving this way were more successful reproductively than were those that behaved differently.' Organisms doubtless are selected for behaving *as though* they are aware of how to maximize their fitness, but this does not require any assumptions as to what is going on inside their heads" (1977, p. 51).

So described, it does not seem to me there is a great difference between sociobiology and old-fashioned behaviorism. Both are deterministic about all human behavior in the strict sense that they see each action as caused by antecedently sufficient conditions, and they both reject any mentalism in the explanation of behavior. The only difference is in the relative weight that they attach to genetic constraints—the behaviorist tends to emphasize the role of operant conditioning in the determination of behavior, and the sociobiologist tends to emphasize the causal role of the genotype. But this still looks to me like a matter of degree. Even the most enthusiastic behaviorist would concede that there are some genetic limits to what conditioning can achieve, and all sociobiologists grant that, within the genetic potential, environment makes a large difference to behavior.

Now, in reply to this explanatory paradigm currently used in sociobiology, I want to propose an alternative approach. This alternative approach can be construed as providing us with an alternative explanatory paradigm for certain of the phenomena under investigation, but the adequacy of the explanatory paradigm has to be tested in actual research practice. I propose that we separate the hypothesis of nonteleological, nonintentionalistic mechanisms of natural selection from the hypothesis of nonteleological, nonintentionalistic explanations of animal behavior. The hypothesis I want to put forward is that the explanation of large areas of human behavior and presumably large areas of other animal behavior is ineliminably intentionalistic in form. Now, this is an empirical hypothesis. It is a claim that the correct causal account of human and animal action requires mention of intentionalistic entities in the causal sequence, which is not eliminable in the way that we eliminated any apparently intentionalistic or teleological elements in the account ascribing plant behavior to purely mechanistic accounts of auxin and in the way that we apparently—although I am not so sure we really did—eliminated any elements of intentionality in the account of bird migration in favor of an account couched solely in terms of the mechanisms of hormonal balances in the hypothalamus. The hypothesis is that intentionality plays a causal role, and, in consequence, any causal account has to mention the causal role of intentionality. But the hypothesis is not that there are no neurophysiological mechanisms underlying intentionality. On the contrary, it seems to me that one of the great neglected questions in sociobiology is, "How did intentionality evolve?" Indeed, as a kind of challenge to the sociobiologist, I would ask, "If intentionality does not play any causal role—if it is just an epiphenomenon—then why did it evolve? Why has it evolved in humans, the higher primates, and probably lots of other animals as well, if it has no adaptive value?"

In order to give some substance to this hypothesis, I want to explain very briefly how intentional explanation works and how it differs from ordinary mechanical forms of causal explanation. Before explaining intentionality, however, I want to call attention to the fact that intentionalistic explanations are the standard form of explanation that we use for most animal behavior already. Consider the following passage, also from Barash, who previously told us that discussions of intentionality were just a manner of speaking in sociobiology. In considering the question of why infant rhesus monkeys who have been separated from their mothers are more bothered—that is, more demanding of attention on being reunited with their mothers—when the original separation has been caused by the mother being removed from the infant than they are when the infant has been removed from the mother, Barash (1977, p. 305) offers the following plausible answer: "Separation caused by departure of the infant is less likely to indicate maternal negligence than is separation caused by departure of the mother. Therefore, infants should be relatively unconcerned about the former, whereas when they have some reason to doubt their mothers' devotedness to them, it would be adaptive for them to behave in a manner that will reduce the likelihood of such separations occurring in the future." Notice that this paragraph is loaded with intentionalistic vocabulary of *doubt, negligence, concern,* and *devotion.* It is not clear how we are supposed to be able to eliminate this vocabulary as we did in our discussion of auxin or even in our discussion of bird migration, where we apparently eliminated any element of intentionality by talking about the hypothal-

amus and the hormones. I want to suggest that there is nothing
wrong with Barash's explanation as it stands, although of course,
as in any causal explanation, we will always need to seek a deeper
causal account in terms of more fundamental principles. That is,
our intentionalistic account is presumably based on neurophysiology, just as neurophysiology is presumably based on biochemistry, and just as biochemistry is, in turn, based on fundamental principles of physics. At least, such is a reasonable working hypothesis
on which to conduct our research. My point is that we make a serious mistake if we ignore the intentionalistic level—and that this
is a mistake is shown by the inadequacy of sociobiological accounts
of communication, a point I shall come to shortly.

Here I want to say a few words about the characteristics of
mental states. We need first to distinguish between those mental
states that are directed at objects and states of affairs in the world
from those that are not. States such as beliefs, hopes, fears, and
desires are in some sense directed at objects and states of affairs.
They have to be about something. For example, a belief must be
a belief *that* such and such, and if I have a fear I must have a fear
of such and such. These differ from such mental states as pains,
tickles, and itches. My pains, tickles, and itches are not *about* anything. They are not directed *at* anything. Many philosophers and
psychologists call this feature of directedness *intentionality.* Intention in the ordinary sense is one kind of intentionality, but belief,
fear, and hope are all intentional in the sense that they are *directed,*
they are *about* something. If this is the case, we will also need to
distinguish between what we might call "the propositional content
of the intentional state" and its "intentional mode." You can see
this difference if you consider a sequence of sentences such as "I
believe you will leave the room," "I fear you will leave the room,"
"I hope you will leave the room," and so on. In each of these, we
have the same content—namely, that "you will leave the room"—
presented in different psychological modes. Notice, furthermore,
that these psychological states, these intentionalistic states, relate
to reality in different ways. We say such things as beliefs are true
or false, and we have evidence or reasons for our beliefs. Our desires or wishes are not true or false in that way, and we do not have
evidence for our desires and wishes. It is, so to speak, the respon-

sibility of the belief to match an independently existing reality, and if we discover that the belief does not match the reality we change the belief. But in the case of the desire, it is the responsibility of the world, so to speak, to match the desire. If the world does not come out the way we want it to or if we do not do the things we intended to do, we cannot patch things up by saying, "Oh well, I guess I just had the wrong intention, the wrong desire." I want to describe that difference by saying that such states as beliefs have the *mind-to-world direction of fit*. It is the responsibility of the mental state to match an independently existing reality, whereas desires and intentions have what we might call the *world-to-mind direction of fit*. It is, so to speak, the world that is wrong if my desires do not pan out, if they are not fulfilled, and if my intentions are not carried out.

The simplest kind of intentionalistic explanation of behavior involves the notion of desire. We explain what an animal does in terms of its desires. But consider the following very simple explanation of human behavior: "Why did the agent put the coin in that machine?" Answer: "It's a Coke machine, and he wants a Coke." Question: "Why does he want a Coke?" Answer: "He's thirsty; he wants to drink the Coke." Or consider the comparable piece of explanation for animal behavior: "Why is that animal moving in that peculiar fashion?" "It's stalking that other animal over there, its prey; it wants to catch the prey, and in order to do that it is trying to avoid being seen." "Why does it want to catch the prey?" "It's hungry; it wants to kill and eat its prey." In discussing these two cases, we first need to distinguish between primary and secondary desires. This distinction is sometimes called the distinction between *ultimate* and *instrumental* or between *basic* and *derived* desires (Nagel, 1970; Pugh, 1977). The desire to put money in the machine, like the desire to catch the animal, is derived from a more basic desire, the desire to drink the Coke or the desire to eat the animal. Derived or secondary desires are perfectly legitimate desires, but they exist only by virtue of primary desires. If the agent loses the primary desire, he will, other things being equal, lose the secondary desire. The logical form of the causal explanation in terms of human desire, in this case, goes as follows. First, the agent has a representation of himself in the desire mode putting a coin in the machine. This desire in turn was caused by a more basic desire, the desire to drink the Coke. This desire consists of a representation of himself drinking the Coke, again in the desire mode. And, again, if the agent acts on his desires, he will realize the conditions of satisfaction of those desires. That is, he will put a coin in the machine, and he will drink the Coke. An exactly parallel account can be given in the animal case. The animal has a desire to catch its prey. This consists of a representation of itself catching its prey in the desiderative mode. If it acts on this desire and succeeds, the animal will then catch the prey. Its desire to catch its prey was derived from a more primary desire, the desire to eat the prey. This desire consists of a representation of itself eating its prey, again in the desiderative mode. If the animal acts on this desire, it will realize the conditions of satisfaction of that representative content; that is to say, it will eat the prey. This form of explanation is very common. It is so common, in fact, that we are almost inclined to overlook it, as it is so pervasive and pretheoretical in our very notion of human behavior and, derivatively, of animal behavior. Nonetheless, it seems to me to have certain very peculiar logical properties that are not common to other forms of causal explanation, and I want to call your attention to three of these properties.

In the first place, the connection between the cause and the effect—that is, between the desire and the action—is not merely a causal connection; it is also a special kind of a logical connection. Indeed, it could not be a causal connection of the intentionalistic kind if it were not also a logical connection of this special kind. The reason it is a logical connection is that the specification of the intentionality of the cause necessarily involves a specification of the effect. The reason for that, as we saw in our discussion of intentionality, is that the cause is a representation of the desired effect. Therefore, it has an internal connection to the desired effect: It could not be *that* representation if it did not represent *that* action. Unlike other sorts of causation, intentional causation necessarily involves a logical connection between the cause and the effect.

Secondly, because the cause and the effect are logically related, the role of causal laws in relating intentional causation to its effects is different from the role of causal laws underlying ordinary mechanical causation. There may indeed be causal laws concerning intentionality, but many of them are tautological or nearly tautological in form. If I give a causal account of the behavior of a falling object or the rise in pressure of a gas that is being heated, in each case I can cite a universal causal law that underlies the particular causal claim. In the one case it is the law of gravity, and in the second case it is the gas laws. But if I am asked to cite a law underlying the causal account I gave of the other two cases, the case of the thirsty man and the hungry animal, the only laws that I can cite would appear to be very implausible as scientific laws. I could say, for example, that people who are thirsty tend to seek something to drink and that animals that are hungry tend to seek something to eat. But part of the trouble with these statements is that the very words *hungry* and *thirsty* already seem to imply the notion of a tendency to behavior. And the reason for that will be obvious from everything I have said. Hunger and thirst are both species of desire, and desires are representations of the states of affairs desired. Because hunger and thirst are logically related to their effects, the laws relating hunger and thirst to their effects are not like ordinary laws. Indeed, I am reluctant to call them *laws* at all.

A third feature of intentionalistic explanations is that they are not deterministic in form. That is, a particular intentionalistic explanation may be deterministic in its content; it may say that, given such-and-such intentional states, the animal could not have done otherwise—it may say that the desires were so powerful and so overwhelming that the animal could not have acted otherwise, all other conditions remaining the same. But that is not a feature of the form of the explanation. As far as the form of the explanation is concerned, it is possible for there to be explanations in terms of desires that are not deterministic. That is, it is consistent to say "I put the money in the Coke machine because I wanted a Coke, and I wanted a Coke because I wanted to drink it, and I drank it because I wanted to drink it" without saying "I couldn't help myself, I couldn't have done otherwise, the desires were so overwhelming as to make it impossible for me to respond differently." It is tempting for us to think of desires as acting like the parallelogram of forces, on agents, so that we think of the resolution of conflicting desires as somehow or other like a vector of forces in Newtonian mechanics. But in fact you will have noticed that, in my discussion of the logical form of intentionalistic explanations, I spoke of the agent *acting on* the desire. The specification of the desire does indeed give a causal account of the action, but in some cases it does not determine the action. The agent determines the action. He decides to act on the desire or not. Because of an obsession with the Newtonian model of explanation, we are inclined to find this very puzzling. We tend to think, for example, that if his desire did not determine his action, then his action must have been irrational or perhaps even random. But of course no such consequence follows. Given a set of conflicting desires, an agent will have reasons for performing or not performing a variety of actions. It is, in general, up to him to select which of these desires he chooses to act on. Sometimes, if he is in the grip of a very powerful lust or drive or passion, he cannot make any such choice, but not all actions, certainly not all human actions, are of this compulsive kind.

The paradox that I find in the sociobiological literature can now be stated. In fact—that is, in actual practice—sociobiologists use intentional forms of explanation for animal behavior all the time. Barash, Wilson, and Pugh (in Pugh's *The Biological Origin of Human Values,* 1977), for example, all use intentionalistic explanations repeatedly, but they seldom give any explicit recognition of the fact that the explanations being offered of animal behavior are frequently intentionalistic, and in consequence there is no recognition of the special logical properties of intentionalistic explanations. And, on those rare occasions, as in the passage from Barash (1977, p. 305), when there *is* an explicit recognition of the intentional character of the explanations, it is passed off as merely a manner of speaking. I believe it is not just a manner of speaking; it is part of the causal account of much animal behavior, at least of the higher animals. The higher animals at least are capable of forming representations of their future actions and of then acting on those representations. One of the many questions this poses for the sociobiologist is, "What sort of neurophysiological mechanisms—and what sort of evolutionary development of those neurophysiological mechanisms—made intentionality possible?" I have not found any discussion of this question in the books by Wilson, and Barash, and Pugh.

Intentionality is an embarrassment not only to sociobiology but also to a lot of philosophers, and I want briefly to mention one attempt the latter have made to get rid of it. Many philosophers believe that although we can say—and, indeed, say truly—that desires cause actions the real causal laws relating our mind to our behavior will be stated not in terms of intentional states but in terms of some more intrinsic characterizations in neurophysiology, perhaps even at the atomic level. On this account, no causal efficacy is actually ascribed to intentionality. It is as if I said, "The event described to me by Bill caused the event described to me by Sally." That is a genuine causal statement, but there is no causal efficacy ascribed to being described by Bill. The actual causal relation between the events will have to be stated in other terms. A similar account is often given of intentionality, when it is said that the causal efficacy will have to be ascribed to some more intrinsic characterization of what the intentional states are: for example, to neurophysiological states in our brains. There is a very simple objection to this attempt to eliminate the causal efficacy of intentional states. Often we have several desires for doing one thing, and we know which of those desires we acted on. I may, for example, have had three reasons for voting for President Carter, but in fact know that Reasons 1 and 2 had no effect on me at all. It is true that I did desire those consequences and that I knew Carter would produce those consequences, but that is not why I voted for him. I voted for him because of Reason 3. Now, in such a case I am aware, in the most direct fashion, of the causal efficacy of my intentional states, and there is no way to eliminate this kind of causal efficacy by saying, "Oh well, it's just a manner of speaking."

In the course of this discussion, I have advanced some fairly strong empirical hypotheses about the explanation of at least some animal behavior. Yet it seems to me that the account I have given is a perfectly ordinary, commonsense account. There is nothing fancy or even particularly adventurous about it. Why, then, does it encounter so much resistance from psychologists and sociobiologists, and why has it received so little explicit attention? I think the answer is that many biologists, along with psychologists, have the illusion that behavior is somehow observable in a way that intentional states are not observable. But in my account there is no way to observe behavior without observing intentional states. To say that an animal is eating, or is stalking prey, or is engaged in aggressive behavior is already to ascribe to it intentional states, as much as when we say the animal has certain beliefs and desires. The actual body movements of the animal could not count as behaviors of the kinds in question unless we were already ascribing beliefs and desires to the animal.

The general theme of my discussion has been that it is essential to introduce the concept of intentionality and of intentionalistic modes of explanation into the conceptual repertoire of sociobiologists in order that they may be able to account for the actual phenomena of animal social behavior. As a further illustration of this point, I want to show how the notion of intentionality is essential to any account of animal communication. I will begin by considering Wilson's account of communication, because it seems to me it is one of the weakest portions of his book and that the weakness of the account derives from the fact that his notion of communication is devoid of any explicitly intentionalistic concepts. Wilson (1975a, p. 176) defines communication as follows: "Biological communication is the action on the part of one organism (or cell) that alters the probability pattern of behavior in another organism (or cell) in a fashion adaptive to either one or both of the participants. By adaptive, I mean that the signaling, or the response, or both, have been genetically programmed to some extent by natural selection." Now this definition contains the notion of a signal, and that notion is subsequently defined as follows (Wilson, 1975a, p. 183): "Let us define a signal as any behavior that conveys information from one individual to another, regardless of whether it serves other functions as well." I think it is difficult to take either of these definitions very seriously, because even in the most sympathetic attempt to read the definitions almost any transfer of information would count as communication. But it is essential that, in giving our account of communication, we should be able to distinguish genuine communication from other forms of biologically mediated transfer of information from one organism to another. The *reductio ad absurdum* of Wilson's whole account comes when he treats the antlers of a male deer as a form of communication of the sexual identity of the deer. To be consistent, he would then have to say that the absence of antlers on a female deer is also a signal on the part of the female; it is an act of communication of her sexual identity. It is easy to see why the definitions force Wilson to take this line; but this line forces him to count almost anything as communication or signaling. If I go for a walk and another animal sees me walking, then my walking is an act of communication, since it conveys information to the other animal, and since I have been genetically programmed to be capable of walking. Furthermore, by this definition the facts that I have two ears, hair on my head, a nose on my face, and am gravitationally attracted to the earth are also forms of communication. The concept of communication is rapidly becoming vacuous, because anything whatever— that is, any perceptually accessible trait that an organism has—is now a signal or an act of communication.

What are the features that genuine cases of communication have that these other forms of information transfer do not have and that we need to capture in any sociobiological account of communication? Very briefly, without going into all the technical details involved, we can state the essential feature of genuine cases of communication as follows: the "speaker" or "sender" of the signal has an intentional state, such as a belief or a fear or a desire, and he or she performs some action with the intention of producing in the "hearer" or the "receiver" of the signal the awareness that he or she has that intentional state by means of getting the receiver to recognize the intention to produce in the receiver the awareness that the sender has the intentional state (Grice, 1957; Searle, 1969). In this account, any genuine act of communication involves a double level of intentionality. First, there must be an intentional state in the speaker or sender of the signal. Second, the speaker or sender must intentionally perform some action designed to induce in the receiver or hearer an awareness of the speaker's intentional state. Notice that when we get away from examples such as the antlers of the deer to genuine cases of animal communication, we find that these features are present. Even for such a "lower" animal as the honeybee, we find it natural to describe the waggle dance as a form of genuine communication and even as a language, because we find it natural to describe the behavior of the honeybee in these intentionalistic terms. That is, the worker bee, having discovered a food source and therefore knowing of the presence of the food source at some distance from the hive, performs the waggle dance. The knowledge of the food source is the bottom level of intentionality; the intentional performance of the dance is the second level. Her fellow workers recognize the features of the waggle dance, and therefore an intentional state corresponding to the bottom level of the intentionality of the speaker is induced in them. They then know where the source of the food is. In this account, we have no intellectual discomfort at all in construing the waggle dance of the honeybee as a case of communication. It is very much like a human speech act. But there is something intellectually uncomfortable about assimilating the antlers of a deer or the color of an animal's legs to a speech act, and the reason for this discomfort is quite clear. All communication involves the intentional transfer of an intentional state from the speaker to the hearer. But we do not have a case of communication in those cases, and they are numerous, where the receiver simply gets an intentional state—that is, where the receiver just acquires an awareness of a piece of information that he or she did not previously have—without any corresponding intentional state in the sender or without any act on the part of the sender that is intentionally designed to convey the intentional state. We just have a case of information transfer or information acquisition.

Making the crucial distinction between communication and other forms of information transfer is not a piece of pedantry unrelated to real empirical facts. On the contrary, genuine communication works on quite special causal principles. A discussion of these special principles necessarily involves the introduction of intentionalistic notions. But we will not get clear about the special features of the intentionality involved in communication unless we get a clear conception of what communication is. The ability to communicate appears to be an essential part of the evolutionary account of the origin of many species of animals, not just humans. We will not get an understanding of these animals or of the evolutionary role that communication has come to play until we get a clear conception of communication; and that conception, I am arguing, essentially involves the notion of intentionality.

I want to conclude this discussion by illustrating some of the

fascinating problems that arise when one takes intentionality seriously as a problem for the sociobiologist. As an example, I will
consider the sociobiological explanation of the incest taboo (Bischof, 1972; I am indebted to Norbert Bischof, Bernard Williams,
and others at the 1977 Dahlem conference in Berlin on Biology
and Morals for discussion of the incest taboo). For the sake of the
argument, I will assume that the sociobiologists are right in claiming that the incest taboo is a genuine cultural universal. That is, I
will assume that, although there are cultural variations in which all
outside the nuclear family are excluded as potential sexual partners, all human cultures exclude potential sexual partners from
within the nuclear family. Sexual relations with one's own parents,
siblings, and children are universally prohibited. Granted that
there are odd cases, such as the Pharoahs, where the universal tendency to prohibition is overcome by *noblesse oblige,* and granted that
plenty of incest actually occurs (there is not much point in having
a prohibition unless there is some temptation to do that which is
prohibited)—still, for the purposes of this discussion we will assume there is a universal prohibition against sexual relations with
blood kin within the nuclear family. It is this taboo that is a candidate for sociobiological explanation.

In this case, as in the other cases, the Darwinian model proceeds by two steps. First, there is the explanation of the evolutionary advantage of the phenomenon (the functional explanation):
The phenotype has survival value or increases "inclusive fitness"
or increases "selectional advantage." In this case, the advantage
comes from recombining the genetic material of the species, thus
producing much greater variety among its members. If incest were
universally required rather than prohibited, we would have all the
disadvantages of biparental reproduction without the genetic advantages. Second, there is a specification of the mechanism by
which this genetic advantage is achieved (the causal explanation).
In this case, leaving aside various complexities, the mechanism is
simply that close personal relationships during the years the child
grows up inhibit sexual desire for the people with whom he or she
grew up. The favorite item of evidence in support of this claim is
the kibbutz syndrome. Children brought up in close proximity in
the collective social arrangements of the kibbutz do not subse-

quently marry or produce children with each other. The typical
kubbutznik does not marry the girl next door, because she was not
sufficiently next door. She was in his nursery, playpen, and sand-
pile from the beginning. Thus the mechanism that inhibits sexual
relations among one's kin is not initially related to social norms; it
is, rather, a natural inhibition produced by extreme familiarity at
certain stages of development. Just as in the bird migration case,
the birds do not migrate in order to increase their genetic fitness—
rather, they have survived, they have genetic fitness, because they
are the sorts of birds that are going to migrate anyhow—so in the
incest taboo case human cultures do not have the incest taboo in
order to increase their genetic fitness—rather, they have survived,
they have genetic fitness, because they are the sorts of creatures
that are going to feel inhibited about sexual relations with their
immediate kin anyhow.

But there are several puzzling features of this pattern of
explanation. In this case, unlike the plant case and to a certain extent unlike the bird case, both the phenomenon to be explained
and the explanatory mechanism are explicitly intentionalistic. The
phenomenon to be explained is, after all, a universal cultural norm.
Its intentional content is "Thou shalt not have sexual relations with
thine immediate blood kin." And the explanation of this prohibition is in terms of an inhibition. Its intentional content is "You will
feel inhibited about having sexual relations with people you were
brought up with." But notice—the intentional content of the inhibition is different from the intentional content of the prohibition,
both in form and in subject matter. The incest taboo is a moral
prohibition, but the mechanism that the sociobiologist explains is
a natural disinclination. How do we get from the latter to the former? That is, supposing that the account of the inhibition is correct
(and it is very loose, by the way; why don't parents seduce their
children on this account, since although the child grew up with the
parent the parent certainly did not grow up with the child?), still
it does not explain the intentional content of the prohibition, or
the peculiarly intense feelings that the prohibition arouses, or even
the need for the prohibition in addition to the inhibition in the
first place. The inhibition makes no reference to blood relatives,

nor does it have any moral content; it simply disinclines the organism to have sexual relations with a certain class of familiar companions. But how do we get from that disinclination to the incest *taboo*? On this question, the sociobiologist is silent. Are we supposed, for example, to hypothesize that the natural disinclination eventually evolved into a prohibition? If so, we would be forced to predict that eventually the Israelis will evolve a moral norm forbidding sexual relations among members of groups that grew up together on kibbutzim. The idea is preposterous, but unless we are prepared to make some such prediction it is hard to see how the identification of the inhibition by itself has any explanatory power at all in accounting for the universal incest taboo, since the intentional contents are so different.

Notice a second feature: Although the intentional content of prohibition is different from the intentional content of the inhibition, it is identical, or nearly so, with the intentional content of the functional explanation. The functional explanation says, "You are better off genetically if you do not commit incest." The moral prohibition says, "You are better off morally if you do not commit incest." Why do they match so closely? One possible answer might be that the moral prohibition is just an epiphenomenon, that the real causal work is done at the level of the functional explanation and the causal mechanisms that enable it to work over long periods of evolutionary time. In this account, the match between the moral prohibition and the functional explanation is just an accident; any old moral prohibition would do just as well, since it does not matter in the actual determination of the behavior. The behavior is actually determined by the disinclination, and the disinclination is a phenotype that survives because it has adaptive value and is determined by some underlying genotype. But it seems this epiphenomenalistic account cannot be right. It runs afoul of the fact that the incest taboo obviously does have an effect on people's behavior, and, indeed, unless the incest taboo functions causally there is nothing for the sociobiologist to explain—only a collective illusion. That is, our problem was to explain the universal incest taboo, but that problem was posed on the assumption that there actually is such a phenomenon—that there is a universal prohibition and that it plays a role in motivating behavior. If we end up with epiphenomenalism, we have denied the premises on which our investigation began.

I make these remarks about the incest taboo not to criticize sociobiology or even the sociobiological account of the incest taboo; rather, my aim is to call attention to some of the problems that arise when one recognizes the crucial role of intentionality in accounting for animal behavior and especially human behavior. The essential insight of sociobiology is that behavior is a phenotype that can be studied from an evolutionary point of view. I entirely agree with this insight, at least as a working hypothesis for further investigation. What I am suggesting in this chapter is that an understanding of the essential character of that phenotype requires an understanding of the role of intentionality in animal behavior. What we need to do is rid the sociobiologist of the illusion that there is something unscientific about introducing the study of mental states as part of the study of animal behavior. On the contrary, I hold the view that it is unscientific to ignore the presence of the intentional states when they quite clearly play such an important causal role in the account of the behavior. I believe the illusion that intentional states are not subject to scientific analysis is a mistake derived from a particularly depressing period in the history of psychology, a period that I hope is now over.

9

Garrett Hardin

Nice Guys Finish Last

$\mathcal{E}$dward O. Wilson is only one of a number of people who have helped create the new synthesizing discipline of sociobiology: The total roster would have to include David Barash, Richard Dawkins, W. D. Hamilton, Maynard Smith, and Robert Trivers. But Wilson's book *Sociobiology: The New Synthesis* (1975a) has, in part because of its lavish scale, attracted the lion's share of attention and criticism, and for that reason the present discussion revolves around Wilson's work and the public reaction to it.

Wilson's *Sociobiology* is a monumental work. And monuments attract graffiti: This one has already been defaced with the epithets *racist, sexist, elitist, biological determinist, Social Darwinist,* and *reductionist.* As Kenneth Boulding (1977, p. 3) has said, "There is something in the 'ist' sound which conveys an almost snakelike hissing and venom." Anyone reading the very voluminous collection of reviews of Wilson's book can sense the venom. At public discussions of sociobiology, the hissing is recordable.

I too have a quarrel with Wilson—a minor one, about terminology. All but one of the twenty-seven chapters in his book do in fact deserve the name *sociobiology*—the sociology of biological organisms (other than human beings). The last chapter, however, together with parts of others, really should be called *biosociology*;

that is, human sociology erected on a foundation of biological principles. Understandably, some sociologists who have spent a lifetime
creating a theoretical structure that is largely devoid of significant
reference to biological principles are loath to admit that a new
foundation is needed. However much the defenders of sociological
tradition may resist the inroads of "biologism" into their subject,
their reaction to the appearance of sociobiology is a casebook study
in territorial behavior. Anthropologists in particular seem to resent
most keenly what they perceive to be academic piracy; for instance,
see Marshall Sahlins' *The Use and Abuse of Biology* (1976b).

Wilson (1975a, p. 4) as much as asked for such criticism by
throwing down the gauntlet in his first chapter: "Taxonomy and
ecology, however, have been reshaped entirely during the past
forty years by integration into neo-Darwinist evolutionary theory—
the 'Modern Synthesis,' as it is called—in which each phenomenon
is weighted for its adaptive significance and then related to the
basic principles of population genetics. It may not be too much to
say that sociology and the other social sciences, as well as the humanities, are the last branches of biology waiting to be included in
the Modern Synthesis. One of the functions of sociobiology, then,
is to formulate the foundations of the social sciences in a way that
draws these subjects into the Modern Synthesis. Whether the social
sciences can be truly biologicized in this fashion remains to be
seen." The modesty of the last sentence has not prevented establishment defenders from perceiving the invasive threat implied in
identifying sociology and the humanities as "the last branches of
biology." Whatever the merits of his program, Wilson has made a
tactical error in announcing it so publicly. Although it may be true
that such statements serve to "rally the troops," it is apparent that
the troops rallied are not all friendly.

Wilson has unnecessarily attracted enemy fire by one other
error, and this one is both tactical and methodological. On the last
page of his book, he says, "The transition from purely phenomenological to fundamental theory in sociology must await a full, neuronal explanation of the human brain. Only when the machinery
can be torn down on paper at the level of the cell and put together
again will the properties of emotion and ethical judgment come
clear" (1975a, p. 575). In point of fact, his book is a convincing

demonstration that the reduction of ethical, social, and political phenomena to the molecular level is not necessary (even if it is possible, which is dubious). Bernard Davis (1976) points out that such a reductionist goal makes sociobiology unnecessarily vulnerable to criticism. Reductionism contributes little or nothing to the understanding of the interactions of organisms. Sociobiology is part of what has been called "skin-out biology," as opposed to a reductionist science such as molecular genetics, which is "skin-in biology." Proper strategy in science calls for a focusing on the appropriate level in the hierarchy of biological organization. An example from the past, from the days before science was self-conscious, should make the point clear. All the major cereal grains were domesticated and greatly improved by preliterate human beings millennia before they knew anything of DNA, chromosomes, or gametes. Plant breeding is skin-out biology. So also is sociology, even in its sociobiology variant. The éclat recently accorded the skin-in subject of molecular genetics should not blind biologists to this fact. The issue of reductionism is a red herring that should be ignored.

What cannot be ignored are political inferences that several scientists of standing draw from sociobiology. Richard Lewontin, supported by his disciples (as Robert Morison, 1976, refers to them) and speaking in the name of the Cambridge-Boston chapter of Science for the People, has inveighed frequently and forcefully against what he perceives to be the implications of the new hybrid discipline. For instance, see the statement by the Sociobiology Study Group (1976), signed by thirty-five persons, including S. Gould, R. Levins, L. Miller, and R. Lewontin. Nicholas Wade (1976), an experienced reporter for *Science,* reviewed the documents and concluded, "In short, the Sociobiology Study Group has systematically distorted Wilson's statements to fit the position it wishes to attack, namely that human social behavior is wholly or almost wholly determined by the genes." The operational words in this summary are "the position it wishes to attack," and this position is that of rigid determinism—*which nobody is defending*. People who think heredity has something to do with behavior do not assert that there is a one-to-one, or even a near one-to-one, correspondence between inherited chemical molecules and observed behavior. The most that is claimed is that inheritance influences behavior—*as does a myriad of environmental factors*. Anyone who would like to effect a statistical change in human behavior has two general options open: (1) to change gene frequencies by selection (either of the classical sort or possibly by some future capability of directly modifying individual genomes—"genetic engineering") or (2) to change the environmental factors in some way. The choice between the two approaches is esentially an economic one, where economics is understood in the broadest sense to include costs of all sorts and to encompass not only present costs but also future costs (which posterity must pay). The decision is never an easy one. There is no widely accepted theory to guide us, but nothing is gained by introducing the false issue of determinism where nothing more than influence is asserted. Determinism is a straw man set up to divert attention from the taboo the antisociobiologists are evidently trying to establish. Given this goal, their tactics in introducing the straw man are well chosen, because it is one of the ineluctable properties of a fully operational taboo that it must not be explicitly acknowledged (Hardin, 1973, Preface).

When a controversy becomes sufficiently bitter, the polarization process inevitably leads some of the participants to make extreme statements that cannot be intellectually supported. Lewontin has gone so far as to say that "Nothing we can know about the genetics of human behavior can have any implications for human society." This remarkable assertion led Roger Masters (1976) to ask, "Are we to believe that the correlation between low IQ and the PKU syndrome has no 'implications for human society' and hence that discovery of dietary therapy for this genetic disorder [is] undesirable?" If Lewontin were untrained in genetics, one might forgive the error, but he is not: He is generally conceded to be one of the most competent genetic theoreticians of our time.

Science for the People is generally regarded as a Marxist organization. Whether the members accept this, I do not know; but reading Lewontin's statements produces a strong feeling of *déja vu* in those of us who remember the defense of the Russian pseudoscientist T. D. Lysenko (see Medvedev, 1969; Joravsky, 1970) by the great English geneticist J. B. S. Haldane (1892–1964) before he ended his long membership in the Communist party. The loyalty of the "true believer" (Hoffer, 1951) can overwhelm the most powerful intellect. It is probably not without significance that Lewontin and Levins have recently come to the defense of Lysenko, more than a decade after Soviet science has rejected him (see Rose and Rose, 1976). The bulk of the rest of the scientific community never accepted his unsupported claims. Lewontin and Levins (Rose and Rose, 1977, p. 59) say, "There is nothing in Marx, Lenin, or Mao that is or that can be in contradiction with the particular physical facts and processes of a particular set of phenomena in the objective world."

A most disturbing aspect of the activities of Science for the People has been the single-minded intensity with which they have attacked sociobiology in the public forum. Science, like horse racing, thrives on difference of opinion, but many observers see something more in this intensity than a mere attempt to get at the truth. Morison (1976) comments, "The basic thesis that certain aspects of nature simply should not be investigated because of the possibility of unfortunate social consequences has not been seen in western Europe since Bruno was executed [in 1600] for his interest in the heliocentric theory." True believers are not now allowed to execute heretics, but they have other techniques for tormenting them. Wilson told Wade, "I have wavered about going to several lectures. There has been clearly prearranged hostile questioning. Perhaps a braver soul would not have been concerned, but I find it intimidating" (personal communication). Because of the increasing mental strain on his family, Wilson has refused many invitations to present his views in public. Society gets the science it selects for: What sort of science do the members of Science for the People want? And who is "the People"? Themselves?

So much for the sociology of science: What about the science of sociobiology itself? At the intellectual level, what are the basic principles or paradigms that are giving trouble? So many hares have been started that what is most essential in sociobiology is not self-evident, but I shall risk focusing on one great principle that (it seems to me) both unifies the subject and accounts for much of the revulsion it elicits. This principle can be put in very simple lan-

guage—and in fact has been, by the baseball player Leo Durocher:
"Nice guys finish last." At the risk of being pedantic, let me explore
this folksy statement in depth.

Many people are revolted by Durocher's statement. Why?
Does it assert a falsehood (in which case revulsion is justified)? Or
is it the fact that the truth is said out loud that causes the shock?
As for the facts, how could things be otherwise? "Nice guys finish
last" is no more than a truism *about any game*. The outcome of a
game played legally is statistically determined by ability operating
within rules. Whoever refuses to take advantage of all that the rules
allow will finish last (on the average). Why should this disturb
anyone?

But it does disturb us, all of us occasionally and some of us
most of the time. Out of the disturbance come such defensive assertions as "The meek shall inherit the earth." There are many
interpretations of this cryptic sentence, and there is truth in some
of them; but to use this assertion as the basis for policy is merely
to create a new game with new rules—under which those who refuse to take advantage of the rules (the new rules) will again come
in last. Recall the game of "giveaway checkers" in which the person
who first *loses* all his or her men is declared *winner*. Changing to
the giveaway rule does not eliminate competition, nor does it prevent more able players from winning over less able. There is strategy even in losing men. There is also strategy in meekness, which
can be pursued very aggressively. Habitual losers always have a
strong desire to change the rules of the game, but when they succeed they generally continue to lose.

Perhaps we should never explicitly say, "Nice guys finish
last," because doing so may create a false impression that ours is
a "dog-eat-dog" world. Yet we know that there is a considerable
amount of altruism in the world—or at any rate something that
looks like altruism. How can this be? Quite simply: If Durocher is
on our side, *we* see no cause for complaint. To us—his teammates—he is nice because it is to his advantage. That it happens
to be to our advantage also is secondary. To put the matter another
way, the nice behavior players show toward their teammates advantages them by way of the group. The indirect gain may or may

not account for the whole of altruism, but it is hard to regard it as irrelevant.

Rational analysis of group and interpersonal behavior always focuses on payoffs, on the reward system. What is selected for? What is rewarded? Only behavior that is consistently selected for will persist. How could it be otherwise? Darwin faced this issue first at the individual level; the passage that follows is equally true if "behavior" is substituted for "structure": "Natural selection will never produce in a being any structure more injurious than beneficial to that being, for natural selection acts solely by and for the good of each. No organ will be formed, as Paley has remarked, for the purpose of causing pain or for doing an injury to its possessor. If a fair balance be struck between the good and evil caused by each part, each will be found on the whole advantageous. After the lapse of time, under changing conditions of life, if any part comes to be injurious, it will be modified, or if it be not so, the being will become extinct, as myriads have become extinct" (Darwin, [1859] 1927, p. 197).

But what about the instances in which the behavior of one species benefits another species—for instance, the pollination of plants by insects, the warning calls of sentinel species of birds associated with mammals, and countless other examples? A Darwinian is adamant in insisting that the serving species always gets something out of the act of service. As Darwin put the matter, "Natural selection cannot possibly produce any modification in a species exclusively for the good of another species, though throughout nature one species incessantly takes advantage of and profits by the structures of others. . . . If it could be proved that any part of the structure of any one species had been formed for the exclusive good of another species, it would annihilate my theory, for such could not have been produced through natural selection" (Darwin, [1859] 1927, pp. 197–198).

Durocher's law could not be illuminated better. Darwin's theory is one of evolution by natural selection. It is noteworthy, as Morse Peckham (1959) has pointed out, that although the acceptance of Darwinian evolutionary theory was very rapid in intellectual circles (being substantially complete by the end of the nineteenth century) the idea of natural selection was dismissed or ignored for the most part. It was too disturbing to admit—and to build the structure of human ethics on—the ideas that nice guys finish last, that winners win, and that there is no payoff for purely altruistic actions.

This is the nettle that the sociobiologists have grasped. For the most part, interspecific altruism presents no problem: Darwin said the last word (although a few fascinating puzzles such as the "cleaning symbioses"—see Limbaugh, 1961—still need more elucidation). The area of continuing strong controversy is that of intraspecific altruism—the sacrifices of a mother for a child, for example, or of one friend for another. And what about the sacrifices a human being may make for an ideal—say, that of universal brotherhood? What have Darwin and Durocher to say about that?

Taking the easiest problem first, it can now be confidently said that altruism among kinfolk has a completely Darwinian explanation. Building on insights by Darwin himself, William D. Hamilton (1972) has given a completely satisfactory explanation of kin altruism in terms of kin selection. Whether this approach gives a complete explanation of kinship and incest rules among human beings—as Dawkins (1976) maintains and as Sahlins (1976b) denies—can be left to the future to decide. In any case, the theory of kin selection is a beautiful and powerful extension of Darwin's theory. The sacrifice of a mother for her children presents no problem to the doctrinaire egoist. If by losing her life the mother saves the lives of more than two of her children, there is a net egoistic gain—although the "ego" in this case is a set of genes. To the extent that behavior is influenced by genes, the genes supporting altruistic sacrifice will be favored by selection if the relative frequency of these genes in the population is thereby increased. No problem.

Suppose, however, that remotely related friends help each other: What then? The kinship of mere friends is close to zero, which means that (by the theory of kin selection) a person would have to benefit a near infinity of friends for altruistic behavior to pay off. Yet friends do help friends: Why?

Simple, says Robert Trivers (1971): It is just a matter of reciprocal altruism—"You scratch my back and I'll scratch yours."

Although his presentation grew out of evolutionary theory, it has no necessary dependence on it: The idea of reciprocity is immensely old. It leads to all sorts of problems about fairness, promises, cheating, and sanctions on which the last word is far from having been said. The mere raising of these questions shows the fundamentally egocentric orientation of the idea of reciprocal altruism. Approaching the issue by an entirely different route, the physiologist Hans Selye (1974, p. 21) has spoken of "altruistic egoism," a term that encapsulates the matter very well.

The Darwins, the Durochers, and the common men and women of the world have no trouble understanding that persistent altruism is never pure, that only altruism that is coupled with egoism can persist. I think it is the intellectual community that resists this commonsense insight most strongly. It is this community that is most firmly oriented toward internationalism, the vision of One World, and the idea of universal brotherhood. I have discussed elsewhere the psychoanalytical reasons for this orientation (Hardin, 1977). I will here discuss the problem in a purely formal, nonpsychoanalytical way.

Society gets the behavior it rewards for. Whatever I do in society produces payoffs through two routes, the directly personal (P) and that which comes back to me as a member of the group (G). The total reward (R) is given by this simple equation (remember that payoffs can be either positive or negative): $R = P + G$. If R is negative, the causative behavior will be extinguished (unless I am a masochist—which is another problem); if R is positive, behavior will be reinforced. If P is negative and G positive, then the absolute value of P must be less than G to produce a reinforcing reward.

Since I am only one of many in the group, my share of the group gain (which share is symbolized by G) is small. This is why it is difficult to get an individual to make sacrifices for the group. I could, for example, give $1,000 to my government—but I do not. In that case, G would be greater than zero, of course; if there are a hundred million taxpayers, one can argue that I would benefit by extra government services worth a thousandth of a cent to me. Even granting that the equilibrating process works that nicely, the total payoff for my altruistic act is only ($-$$1,000.00 + $0.00001), which—for all practical egoistic purposes—is $1,000 down the drain. So I never *give* money to the government. Neither, if I am a realist, do I ask or expect anyone else to do so. As the political scientist William Ophuls (1977) points out, never in the history of the world has there been a society that depended on voluntary taxation. Ophuls says, further, "Some aspire to do away with power politics and state coercion entirely by making men so virtuous that they will do what is in the common interest. In fact, this is precisely what Rousseau proposes: small, self-sufficient, frugal, intimate communities inculcating civic virtue so thoroughly that citizens become the 'general will' incarnate. However, this merely changes the locus of coercion from outside to inside—the job of law enforcement is handed over to the internal police force of the superego— and many liberals (for example, Popper, 1966) would argue that this kind of ideological or psychological coercion is far worse than overt controls on behavior" (Ophuls, 1977, p. 150).

In any case, the experience of centuries unambiguously shows that such inner controls suffice for governing only the smallest communities (Bullock and Baden, 1977). Large groups—and, in this context, any community of more than a few hundred is large—must govern themselves by laws based on policy. The cardinal rule of policy is *Never ask a person to act against his own self-interest* (Hardin, 1977). How, then, do we manage to achieve group goals? Necessarily, by coercion. To achieve the fairness that is necessary for political stability, we must have *mutual coercion, mutually agreed on* (Hardin, 1968). For a large group, there is no better way. Adhering to the cardinal rule, the wise man does not try to get the group's work done by appealing to people's generosity or castigating their greed. As Helvétius said in 1758 (Harris, 1974), "It is solely through good laws that one can form virtuous men. Thus the whole art of the legislator consists of forcing men, by the sentiment of self-love, to be always just to one another."

An enduring social order must be built on egoistic impulses, harnessed by mutual coercion, and mutually agreed on (for reasons that are egoistic but not ego centered). I suspect there will always be many people whose minds cannot tolerate so understandable, so logically mechanistic a world as this. These people hope for a world in which the meek shall inherit, a world governed by the principle expressed by Karl Marx ([1875] 1972) of "From each according to his ability, to each according to his needs." But who is to determine the "needs"? If each person determines his or her own egoistic needs, greed will soon make a shambles of social organization, as nice guys finish last. If the group makes the determination and if the group is large, the determining function must be turned over to bureaucrats—and there will soon be complaints of injustice.

Marxists are quite right in being revolted by what they perceive to be the implications of sociobiology: It threatens their value system. What sociobiologists propose to do is explain human actions, in a statistical way, on the basis of genetic *and other natural forces*. But to regard the human conditions as influenced by—one must not say determined by—natural forces is to adopt the tragic view of life, a view that is incompatible with Marxist doctrine. As Morison (1976, pp. 18–19) says,

> It is interesting to reflect that those prone to interpret the human condition in terms of genetic or natural forces at least have the decency to cast their remarks in a mood of tragic inevitability, as Freud (1962) did, for example, in *Civilization and Its Discontents*, or the Greeks in their accounts of the misfortunes of the House of Atreus, or as Wilson himself quotes Arjuna in the *Bhagavadgita*. Those addicted to environmental explanations, on the other hand, may find it too easy to rejoice in the power to achieve equality and eliminate sin by adjusting the external variables, while conveniently forgetting that the same power can emerge less happily in the use of the poison pen, the threatening telephone call, the midnight knock on the door, all delivered according to that random schedule which the operant conditioners have found most effective.

Those who try to force the cooperation of others without recourse to the mutual coercion incorporated into democratically enacted laws pursue a strategy that Reinhold Niebuhr (1937, p. 139) has laid bare: "The self is tempted to hide its desire to dominate behind its pretended devotion to the world. All mature conduct is therefore infected with an element of dishonesty and insincerity. The lie is always intimately related to the sin of egoism."

The difficulty of obtaining a consensus on the limits of altruism in human intercourse is inextricably connected with the fact that every statement about altruism and egoism serves—well or ill—egoistic goals. Morality aside, the optimum strategy for the unabashed egoist is unwavering praise of altruism. This is the central thread of the strategy followed by every demagogue (*demos*, remember, means "people"). Science for the People does not deviate from this strategy.

Looking at sociobiologists with a sociobiological eye, we see that they are in a uniquely vulnerable position. Professionally, they wish to know the truth about individual differences, sexual differences, racial differences, and species differences; to know the extent to which differences in endowment can influence function; and to know the total cost of escaping some of the limitations of nature by intervening in the environment. At every stage in the progress of learning, knowledge is incomplete. Unfriendly bystanders may infer from research reports more than the researchers imply. (Researchers also may make mistaken deductions.) Taboos will be broken, but critics will fail to point this out because of the twofold nature of taboo—to suppress discussion *and* to suppress knowledge of the suppression. A taboo "is a sort of Chinese egg. Inside is the primary taboo, surrounding a thing that must not be discussed; around this is the secondary taboo, a taboo against even acknowledging the existence of the primary taboo" (Hardin, 1973, p. xi).

The sanctions inflicted on the breakers of taboos are as severe in academia as they are in the agora. As economist Paul Samuelson (Wade, 1976) has pointed out, "to survive in the jungle of intellectuals the sociobiologist had best tread softly in the zones of race and sex." The wall of the house of sociobiology is decorated with a sampler that bears the motto "Nice Guys Finish Last." It will be ironical if the future of the social sciences is determined by this law reflexively acting on science itself to favor hypocrisy over candor and the earnest search for truth in the intellectual jungle of academia. This disaster need not occur if the scientific community is sufficiently conscious of what it is doing and if it religiously adheres to Milton's high standard of freedom of inquiry.

10

Joseph S. Alper

Ethical and Social Implications

*I*n the light of some currently prevailing views about the nature of scientific theories, the title of this chapter is a contradiction in terms. Sociobiology, according to its practitioners, is a science like chemistry and physics, and as such it can be characterized by the same tenets that are applicable to all sciences. A scientific theory is said to be objective; its content is independent of the investigator and of the particular environment in which he or she works and lives. The scientific method by which the theory is produced is also objective; the logical argumentation and the evidence presented are either valid or invalid, *sub specie aeternitatis*. Consequently, a scientific theory is value free. It entails no ethical or political implications, since values have no objective reality but are rooted in human culture.

The naiveté of these views, although not widely recognized

I would like to thank the members of the Sociobiology Study Group of Science for the People, especially Jon Beckwith, Stephen Gould, Robert Lange, Lila Leibowitz, and Freda Salzman, for their constructive criticism of this chapter.

by scientists, is becoming increasingly well known, so that just a few observations are necessary here. The form of scientific inquiry is determined to a great extent by the dominant forces of society (Beckwith and Miller, 1976). The problems studied and the direction research takes to solve these problems are controlled to a greater or lesser degree by the funding of projects by the government and by large corporations (Noble, 1977). Furthermore, very often scientific and technical research is of a type that can be used either directly or indirectly for the purpose of exerting economic or social control over great numbers of people (Braverman, 1974). The standards for accepting the logic of and the evidence for a new theory reflect not some absolute standard but rather the willingness of scientists to accept a new theory, which in turn depends on the political and social, as well as the intellectual, conditions of the time (Kuhn, 1970; Feyerabend, 1975).

Sociobiology, a recent offshoot of biology that attempts to discover the biological basis of animal and human social behavior, is a theory that not only illustrates graphically the lack of objectivity of science but that also contains very serious social and political implications. The major conclusion of sociobiology seems to be that the social behavior of people, including such traits as sex roles, aggression, territoriality, and altruism, are largely determined by our genes and by evolution. If these characteristics of our society are biologically determined, then the present structure of society is the outcome of natural process. Thus the status quo is justified by an argument from nature. It follows that there is not only little point but also possibly grave dangers in trying to overcome sexism and aggression or in trying to change the present economic and social structure, since the present society is based on natural selection and other biological principles.

As sociobiologists themselves have pointed out, there is no strict logical deductive chain from the hypothesis that traits are biologically determined to the conclusion that these traits are unchangeable. A trait is said to be biologically determined if it is one that remains relatively constant in all known environments. However, there always exists the possibility that new environments could be provided in which these traits would be modified. This limitation on the causal chain is, however, quite subtle. Very few nongeneticists appreciate the distinction between a biologically determined and an immutable trait. In our society, for all practical purposes, the distinction does not exist. For this reason, it is particularly alarming that sociobiological theory, which has no proof that the social behavioral traits it discusses are biologically determined, is taught in high school curricula. The Education Development Center's text *Exploring Human Nature* (1973) presents sociobiological reasoning in such a way as to inculcate in its readers a belief in the inevitability of the status quo.

Sociobiology has its origins in the study of the social behavior of animals. The complex social behavior of bees, ants, lions, and monkeys is familiar to most of us, and sociobiology attempts to account for this behavior in terms of biological principles. This field gained widespread recognition with the publication of E. O. Wilson's text *Sociobiology: The New Synthesis* (1975a). This book seeks to determine general principles and then apply them to all social animals, including humans. Since human beings form a species of animals, they follow the same biological laws. Consequently, human social and behavioral traits can be seen as universals (common to all people), and these traits can be explained in terms of the same biological principles that apply to other animal species. Wilson (1975a, p. 5) states that "The principal goal of a general theory of sociobiology should be an ability to predict features of social organization from a knowledge of these population parameters combined with information on the behavioral traits imposed by the genetic constitution of the species." The "population parameters" referred to are demography and the genetic structure of populations. This program seems to disregard culture, ignores philosophical questions, and seeks to subsume under biology such disciplines as anthropology, economics, and sociology in the "new synthesis."

Sociobiological theory, as applied to the human species, operates in the following manner. It is postulated that differences in human behavioral traits from one person to another are caused by genetic differences between the individuals. If one then assumes that these traits are adaptive—that is, ensure greater reproductive success to their bearers—then these biologically determined traits can be passed on from one generation to the next. In the course of time, those traits providing the greatest adaptive value will dominate and become universal biologically determined traits. Since many of the traits involved are social ones that do not benefit the single individual, it is further postulated that a trait will also prevail if the individual's relatives, who carry some of the same genes, are benefited by that trait (kin selection).

One of the examples Wilson provides will illustrate the operation of the theory. Homosexuality can be explained using sociobiological concepts. It is assumed that homosexuality is genetic in origin. If it is also assumed that homosexual members of primitive societies functioned very efficiently in assisting close relatives (some of whom also carried the "homosexuality genes"), then these relatives might have become more reproductively fit—that is, have had more children—and thus the trait of homosexuality would be preserved and propagated. Needless to say, there is no evidence for any of these assumptions.

This example of the explanation of homosexuality is indicative of the untestable nature of sociobiology. The assumptions on which this explanation is based are unprovable, since they rely on guesses about early human societies. Even if the assumptions were provable, the possibility of alternative explanations is not ruled out. Why could not homosexuality be caused by environmental factors arising in the course of an individual's development? Wilson himself, in a section of his book on scientific methodology (1975a, p. 29) refers to this error of not ruling out alternative explanations as the well-known fallacy of "affirming the consequent." Moreover, by using the mechanisms of individual and kin selection together with those of reciprocal altruism (one person helps an unrelated person in the expectation that the receiver of help will eventually reciprocate) and spite (behavior that benefits neither of the interacting individuals), any social behavior at all can be explained—that is, shown to be adaptive. Unlike a science such as astronomy, in which speculation also runs wild but in which most of the theories are eventually found to be false as a result of new observations, for sociobiology no phenomena could arise that sociobiologists would be at a loss to explain in terms of their theory. Sociobiology has more in common with an all-explanatory system such as Freudianism than with any of the empirical sciences.

It is useful in beginning a critique of the specific postulates of sociobiology to look at the particular behavioral traits and social conditions the theory attempts to explain. Some of the traits and conditions that Wilson cites include indoctrinability, the nuclear family, male dominance and superiority, blind faith, genocide, warfare, xenophobia, and finally the unequal distribution of wealth and prestige. In the first place, there is much sociological and anthropological evidence showing that many of these behavioral patterns are in fact not human universals (Sahlins, 1976b). Second, Wilson never explains why these particular traits were chosen or whether there might be other human social traits of equal or more fundamental importance. Wilson seems to have fixed on them because they are the ones most noticeable in our present society, but it is by no means certain that sociobiologists from a different culture at a different time might not have focused on an entirely different set of traits having little in common with those our sociobiologists choose to study. It is evident that the traits chosen characterize our particular society and that an "objective" theory that will account for these behavioral patterns would go a long way toward justifying that society.

In choosing human behavioral traits that are to be deemed universal, sociobiologists find it useful to make connections with behavioral traits exhibited by animals. Such a connection would constitute evidence that the analogous traits were genetic and also that they arose in humans as a result of evolutionary natural selection. The methodology used in making the connection, however, gives the appearance of mutual reinforcement. First, animal behavior is described in human terms. Examples abound: The word *slavery* is used to describe certain behavioral patterns in ants; in describing "courtship" behavior in animals, such terms as *salesmanship, sales resistance, coyness,* and *machismo* are employed; and such elements of behavior as barter, division of labor, and role playing are attributed to animals. Sociobiologists maintain that such terms, when applied to animals, are being used in a special technical sense and that they are not meant to imply any relationships to human behavior. However, this distinction is easily forgotten, and in fact sociobiologists then use this animal behavior, which they have already characterized in human terms, to argue that the corresponding behavior in humans has its roots in animal behavior and is therefore biological in origin. Once again we should note that these universal traits, which are now attributed to animals, seem to be those types of behavior relevant to the society of the researchers themselves. It is no wonder that ethologists are able to explain competitive and sexist traits in people on the basis of studies of animal behavior; the particular human traits of interest to the ethologists were attributed to the animals by the ethologists themselves.

Once the particular traits to be explained have been singled out (by a highly subjective, value-laden procedure, as we have seen), the sociobiologists must then show that these traits are biologically determined—that is, differences in these traits are due to genetic and not to environmental differences—and that these traits remain unchanged unless there are drastic environmental changes. There is no direct evidence for the genetic basis of human behavioral traits, and in fact it is probably impossible to find evidence either for or against the principle. It is impossible to perform controlled experiments on people, since we cannot, for obvious moral and practical reasons, manipulate the environmental and genetic factors affecting people's lives. Another difficulty lies in the fact that human behavioral traits are not objective realities but are rather constructs created by social theorists. It is not axiomatic that there are genes responsible for the expression of such reified traits. Calling such elements "organs" of behavior (Wilson, 1975a, p. 22) does not provide them with any greater degree of reality. Once again, nonscientific presuppositions of the investigator play an important role in determining the shape of a scientific theory.

In all fairness, it should be stated that Wilson does not consider his theory to be a biologically deterministic one but maintains that environmental forces, including culture, could play a large role. Sociobiology, however, when applied to humans would be meaningless as a discipline if it were not basically biological. The avowed program of sociobiology to base the study of social behavior on the sciences of population genetics and evolutionary ecology precludes, or at least relegates to a subordinate role, the introduction of cultural factors. Inclusion of these factors would require an independent theory of the development of social structures that is, of course, absent from present sociobiological theorizing. This absence is probably unavoidable, since there could be no "new synthesis" in which human sociobiology is incorporated into animal sociobiology, and there could be no withering away of anthropology and sociology unless the same biological principles of animal sociobiology can be applied to the human species without recourse to additional cultural hypotheses.

Sociobiology's major flaw seems to be that it fails to recognize that the human species develops by means of two forms of evolution. Biological evolution ensures survival to the descendants of those animals who were more fit reproductively: in other words, those who adapted better to the particular environment in which they were living at the particular time they lived. The traits that are passed to their descendants are those traits with which these better-adapted animals were born. Biological evolution can be contrasted with cultural evolution, in which acquired traits of one generation are passed directly to the next generation. This latter type of evolution is of overwhelming importance to the human species. It is a much more reasonable hypothesis, with at least as much evidence in favor of it, that human social behavioral traits are not biological universals passed on by Darwinian evolution but rather are cultural traits transmitted by cultural evolution. This latter type of evolution is most directly due to the development of language, via which people are able to react to their environment, to change both it and themselves, and to pass the knowledge of these changes directly to the next generation. The Lamarckian character of cultural evolution enables human beings to adapt very rapidly to changes in the environment—much more rapidly than those species that must rely on biological evolution—and has been of tremendous survival value for the human species.

Cultural transmission of social behavior does not preclude recognition of a genetic component in all human behavioral traits. But society is not simply the sum of the behavior of individuals, as the sociobiologists seem to assume. It is rather the product of complex interactions among people and the institutions of society that have arisen in the course of history. People created society, which in turn forms the political and social environment for their lives. This environment influences their behavior, which in turn can modify society. People always retain the potential for overcoming their environment, revolutionizing social and political institutions, and changing their own nature.

In its attempt to explain all of human behavior in biological terms, sociobiology ignores these complex interactions between genetics and the environment, including society, and assumes that human nature is biologically determined and basically unchangeable. This theory, although presenting itself in the form of objective science, is clearly a product of the political and social climate in which it arose. Both its content and the method used to validate it have been structured to fulfill certain preconceptions, including the justification of the status quo and the rationalization of policies maintaining sexual, social, economic, and political inequality.

We now shall examine some of the specific implications of sociobiology. We recall that the social and behavioral traits deemed basic by sociobiologists are claimed to be universal, largely biologically controlled, and consequently natural. Several of the traits that are repeatedly stressed—namely competition, entrepreneurship, and aggressive behavior—appear to be especially relevant to Western capitalist societies. Wilson believes that these traits are universal because they arise out of the universal conditions of economic scarcity. "The members of human societies sometimes cooperate closely in insectan fashion, but more frequently they compete for the limited resources allocated to their role sector. The best and most entrepreneurial of the role actors usually gain a disproportionate share of the rewards, while the least successful are displaced to other, less desirable positions" (Wilson, 1975a, p. 554).

The use of the derogatory word *insectan* to characterize societies where cooperation rather than competition is the norm is probably the most telling betrayal of Wilson's supposed scientific neutrality. Moreover, his statement is misleading, since scores of societies have been studied that are not characterized by a hierarchy of social positions and an inequality of shares (see, for example, Birket-Smith, 1959; Harris, 1968, pp. 369–391; and other references cited in Sociobiology Study Group, 1977). The view that these behavioral patterns characteristic of capitalist societies are universal over time and in all cultures cannot be substantiated.

Rather than being universals, these traits are quite specific to our culture and as such play a large role in forming the ideology of our society. Ideologies both conceal and reveal the underlying assumptions of the structure of a society and function to legitimize it. It has been pointed out that traditions that at one time served this legitimizing function (for example, religion) lost their efficacy at about the time of the French Revolution and that subsequently, with the rise of positivist schools of thought, more scientific modes of explanation were required for that purpose (Habermas, 1970, pp. 97–99; Gouldner, 1976, chap. 2). Sociobiology, which sees itself as the science of social behavior, seems ideally suited to provide a credible foundation for the ideology of capitalism. This role of sociobiology in supporting the ideology of capitalism and thus of the status quo is rather abstract and cannot be explored fully in this chapter. I turn instead to those implications that are of more immediate consequence.

Whether for good or for bad, our society relies on the opinions of certified scientific experts for making social and political decisions. The testimony of psychologists that eastern and southern Europeans are innately less intelligent than western Europeans was instrumental in the passage of the highly restrictive Johnson-Lodge Immigration Act of 1924 (Kamin, 1974). On the other hand, the opinion of psychologists and sociologists that segregated schools are inherently unequal to desegregated ones influenced the U.S. Supreme Court in the school desegregation case (*Brown* v. *Board of Education*, 1954, see especially fn. 5). If sociobiology is regarded as a science, its proclamations will be taken very seriously, even more seriously than those of the "softer" sciences of psychology and sociology. With these thoughts in mind, we consider the possible effects of sociobiological theorizing on the problems of sexism and racism.

The issue of the role of women in our society has become increasingly important in recent years. Many women have begun to see themselves as an oppressed group, discriminated against in education and in employment opportunities and advancement, and as having been relegated to an inferior social status by the necessity of remaining at home taking care of the house and their children. This raised consciousness has resulted in the creation of women's organizations fighting for equal treatment. On the other hand, because of high levels of unemployment, there is increasing fear of the prospect of a new group of people competing on equal terms for jobs, especially for the relatively few available professional and managerial positions. As a result of these economic conditions, as well as other political and social factors that are beyond the scope of this chapter, the women's movement, after a period of some gains, has met with increasing resistance in the late 1970s. The loss of momentum in the effort to pass the Equal Rights Amendment, various U.S. Supreme Court decisions involving pregnancy benefits and abortions, and the increasing disparity of wages between men and women are only some of the manifestations of this trend. These reverses have led to an increasing intensity of debate as women have become more insistent in their demands.

Such debate, of course, is not a new phenomenon, and during periods of the waxing of women's movements—as for example, the suffragette movement in this country in the early twentieth century—arguments have always been made that women are naturally (in current language, biologically), both physically and mentally, the weaker sex and so cannot expect to play the same role in society as men. Not surprisingly, various scientific evidence and scientific theories have recently been presented purporting to prove the unsuitability of women for positions of responsibility and authority and to demonstrate that women's place is in the home (Hutt, 1972; Goldberg, 1973; Tiger and Shepher, 1975). Of these theories, sociobiology is the most encompassing and sophisticated. The coincidence of the appearance of sociobiological theories rationalizing the retreat from the goals of equality is so striking that even *Time* magazine (1977, pp. 54–63), in its generally favorable article on sociobiology, entitled "Why You Do What You Do—Sociobiology: A New Theory of Behavior," by Ruth Mehrtens Galvin, felt compelled to refer to it.

The importance of sex differences in the theory of sociobiology cannot be overemphasized. In his book, David Barash (1977, p. 283) writes, "Sociobiology relies heavily upon the biology of male-female differences and upon the adaptive behavioral differences that have evolved accordingly. Ironically, Mother Nature appears to be a sexist, at least where nonhuman animals are concerned. There may or may not be similar biological underpinnings of sexism in human societies; we do not know." Wilson (1975b, pp. 49–50) maintains that "even with identical education and equal access to all professions, men are more likely to play a disproportionate role in political life and science. . . . My own guess," he states, "is that the genetic bias is intense enough to cause a substantial division of labor even in the most free and egalitarian of future societies." Both authors stress that they are talking merely about differences and that they are not implying that one sex is better. Wilson adds that even if his guess is correct it "could not be used to argue for anything less than sex-blind admission and free personal choice." Nevertheless, such disclaimers ring false in a society such as ours, where differences have always entailed prejudices. If women are by nature less fitted for the occupations Wilson mentions, which are among the most prestigious in our society, it is difficult to see how women could not be seen as inferior. The combination of a history of centuries of sexual discrimination, together with a scientific theory purporting to explain in biological terms the stereotypes on which this discrimination rests, can only act to reinforce existing prejudices.

We have discussed elsewhere the crucial role that sexual differentiation plays in all aspects of the theory of sociobiology, so only one phase of the argument will be given here (Alper, Beckwith, and Miller, 1978). The fundamental traits on which the theory is based and from which nearly all of the other traits can be derived—namely,"aggressive dominance systems with males dominant over females; scaling of responses, especially in aggressive interactions; pronounced maternal care; pronounced socialization of young; and matrilinear organization" (Wilson, 1975a, p. 552)—are all related to sex. It is the supposed biologically based universal aggressiveness and dominance of males that lead Wilson to his conclusion that women naturally stay at home and men go out to hunt or to perform the modern equivalent. The arguments supporting the assertions of the universality and biologically based sex-differentiated nature of these traits depend on analogies to primate behavior, anthropological studies of different cultures, physiological research on differences in sex hormone levels and on brain differentiation, and psychological studies on the development of the sex-stereotypic behavioral differences in children. Much of this research, especially the use of this research by sociobiologists, is open to serious question and has been subjected to detailed criticism (Chasin, 1977; Salzman, 1977). At the present time, the most charitable position on the sociobiologists' view of women is that it is unproven.

There seems to be general agreement among both the opponents and proponents of sociobiology that sexism, or at least the doctrine that the biological differences between the sexes results in different types of social behavior, is an integral part of the theory. The disagreement lies in the implications of the theory and the question of the responsibility of the sociobiologists. Are sociobiologists responsible if, despite their statements that their theories describe "what is" and not "what ought to be," sociobiology is used as a scientific theory to justify unequal pay or to deny a woman an executive position in some corporation? Since we do not live in an ideal society where people are treated as equals even though they exhibit differences, the realization that sociobiology is only a speculative theory and that even the evidence on which it relies is uncertain seems to be the decisive factor in answering the question.

The issue of the racist implications of sociobiology is a much more delicate one; we are all much more aware of racism than of sexism. Throughout its history, the United States has been plagued by a high degree of racial violence. Our society has made efforts to overcome racism, and academics in particular are usually on their best behavior when discussing this issue. We can imagine the uproar if the passage just quoted on representation in the professions distinguished between races rather than sexes. There can be no doubt that, no matter what qualifications and caveats accompanied the statement, the author of such a statement would be labeled a virulent racist. In the language of the U.S. Supreme Court, race is a suspect classification.

From the time of the publication of Wilson's book, discussions concerning the link, if any, between sociobiology and racism have generated more heat than light. The early statements of the Sociobiology Study Group (SSG) of Science for the People were interpreted to mean that sociobiology was a racist theory, equivalent to Social Darwinism, and some people in the group felt the

need to state publicly that SSG does not in fact believe that sociobiologists are racists. SSG believes that sociobiology is not racist in
the usual sense of the term. However, built into the theory is the
biological inevitability of intergroup conflict. Given the history of
our country and present political realities, such a theory can be
easily used to support doctrines of discrimination. We believe that
sociobiologists cannot divorce themselves from the racial implications of their theory.

A race can be defined as "a breeding population characterized by frequencies of a collection of inherited traits that differ
from those of other populations of the same species" (Goldsby,
1977, p. 21). Goldsby notes (p. 25) that we can replace "frequency
of a collection of inherited traits" with the more precise term "gene
frequencies." It is important to emphasize that the concept of race
is not completely well defined, since there is a great deal of overlapping of the races as defined and since there is no unambiguous
way of dividing a species into groups according to the frequencies
in the appearance of specific genes. In fact, it is questionable
whether the concept of race is a useful one at all.

Modern racism is the doctrine that some races are biologically superior to others. At first sight, sociobiology appears to be
an antiracist theory, since it places such emphasis on the universality of human traits. However, because sociobiology is based on
evolutionary theory, genetic differences among people are crucial.
Evolution is possible because some genetic differences give rise to
phenotypes that have greater adaptive value than others. Bearers
of these phenotypes become proportionally more numerous in suc-
ceeding generations. We have discussed this process in terms of
individual and kin selection. In order to explain such phenomena
as warfare, sociobiologists invoke another form of selection, group
selection. Wilson (1975a, p. 8) defines a group as "a set of organisms belonging to the same species that remain together for any
period of time while interacting with one another to a much
greater degree than with other conspecific organisms. The word
group is thus used with the greatest flexibility to designate any ag-
gregation or kind of society or subset of a society." People in a
group are characterized by a set of gene frequencies that differ
from the gene frequencies of the people in a neighboring group.

In a completely analogous mechanism to that of kin selection, people in one group will tend to cooperate in struggle against those in the neighboring group in order to increase the chances of their genotypes prevailing. Group selection thus explains xenophobia and warfare. In Wilson's words, "Outsiders are almost always a source of tension" (1975a, p. 286). War is the natural outcome of the struggle of one group against another. "*Deus vult* was the rallying cry of the First Crusade. God wills it, but the summed Darwinian fitness of the tribe was the ultimate if unrecognized beneficiary" (1975a, p. 561).

Although Wilson rarely if ever uses the word *race* in his text, it appears that race, as defined by Goldsby, is one type of Wilsonian group. It is true that much of the technical biological literature restricts the term *group* to a subset of a population that usually would be much smaller than what would normally be considered to be a race (Maynard Smith, 1976). Nevertheless, it is clear from the quotation referring to the First Crusade and from his definition of a group that Wilson's concept is much broader than the technical one. Once we have recognized that a race is simply a type of group, we see that for sociobiology racial conflict is natural.

Up to this point in the argument, there is no indication that there are actual racial differences or that one race is superior to another. However, even at this stage the theory has serious implications. Although no justification has been given for the doctrines of racism, racist behavior and prejudice are seen to be natural biological traits that are possibly inevitable and that may even have adaptive value. Barash (1977, p. 311) has written, "Given the evolution of differences and the fact that such differences have been associated with greater genetic similarity within groups than between groups, it seems plausible that a degree of antagonism, the converse of altruism, would occur when representatives of different groups meet. . . . Clearly this suggestion of a possible evolutionary basis for human racial prejudice is not intended to legitimize it, just to indicate why it may occur." Wilson (1975a, p. 286), after describing manifestations of xenophobia, states, "Efforts are then made to reduce them [enemy groups] to subhuman status, so that they can be treated without conscience. They are the gooks, the wogs, the krauts, the commies—not like us, another subspecies surely, a force remorselessly dedicated to our destruction who must be met with equal ruthlessness if we are to survive. . . . At this level of 'gut feeling,' the mental processes of a human being and of a rhesus monkey may well be neurophysiologically homologous." This quotation is striking in that a key aspect of racism, the treatment of a person of a different race as a subhuman, is presented as a biological universal that can be traced to a similarity in the "wiring" of the brains of people and monkeys.

The genetic mechanism for group conflict as described by Wilson is only a hypothesis that is still the subject of controversy among biologists (Maynard Smith, 1976). Moreover, irrespective of the merits of any of the parties to the biological debate, it would seem that the obvious social, political, and economic causes of racial conflict and war are vastly more important than genetic factors (Sahlins, 1976b).

We believe that there is a deeper connection between sociobiology and racism than just the alleged biological nature of intergroup prejudice. Wilson himself goes further in his book (1975a, p. 550): "There is a need for a discipline of anthropological genetics. . . . Variation in the rules of human cultures, no matter how slight, might provide clues to underlying genetic differences, particularly when it is correlated with variation in behavioral traits known to be heritable." The implications of this statement are frightening. The entire race-IQ issue was built on precisely this attempt, to show that the differences in measured IQ between blacks and whites are due to differences in their genes. The history of that controversy has shown that it is probably impossible to measure the heritability of *any* complex social behavioral trait because of the problems encountered in heritability studies (Kamin, 1974; Block and Dworkin, 1976). A by no means exhaustive list of the difficulties includes (1) the definition and quantification of a complex trait such as intelligence (What does IQ have to do with intelligence?), (2) the flaws (including fraud) in the identical twin studies designed to measure heritability, (3) the impossibility of measuring the separate genetic and environmental components of a trait in cases where the genetic-environmental interactions are strong, (4) the methodological error of using heritability estimates for one group to estimate heritability for another group, and (5) the error of assuming that heritability within a group can tell us anything about the causes of differences between groups.

Furthermore, as Wilson himself indicates, genetic variation between groups is small. Lewontin (1972) has shown that, for blood types and presumably for other traits as well, most of the genetic variation in the human species occurs *within* groups. Cavalli-Sforza (1974) has pointed out that most of the small intergroup differences are literally superficial, that is, skin deep. Most racial differences are due to the adaptation of different groups to their different physical environments (climate, amount of ultraviolet radiation from the sun). Given the number of important unsolved problems in biology, it seems peculiar to worry about "slight" genetic differences and to postulate unprovable biological explanations for social phenomena that might arise from these differences. Arthur Jensen (1969) was explicit in his discussion of the biological origin of IQ differences. We might ask, why fund compensatory educational programs if they are destined to fail?

Once the idea of biologically caused differences in behavior is accepted, it is only a small step to a belief in racial stereotypes and then to a judgment of which of the alternative behavioral patterns is superior. Throughout history, the dominant group in the society—whether racial, political, economic, social, or religious—has defined the differences and decided which characteristics are to be deemed inferior. In the course of time, these determinations are solidified and are perpetuated by the institutions of the society.

This final step of defining racial differences and determining which characteristics are the superior ones is, of course, absent from sociobiology. But sociobiological research is not conducted in isolation; its social context is critical. Racism is virulent in this country, and it will, like sexism, probably intensify as competition for scarce jobs becomes even more keen. Just as every discovery in molecular biology is examined for its potential for curing cancer, so any biologically based social theory is examined for its possible political and social implications. For these reasons, we see an intimate connection between the ideas of group conflict in sociobiology and racist ideology that all of us, including sociobiologists, abhor.

The most striking example of sociobiology's destructive potential is without doubt the film *Sociobiology: Doing What Comes Naturally* (1976). This film, produced by a private company for distribution in high schools and colleges, concentrates on the biological nature of behavioral differences between the sexes and on aggressive behavior. One horrifying segment shows American planes bombing Vietnamese villages while the sound track suggests that war is the process by which the better genes win out in the struggle for survival and that the pillage and rape that accompany conquest have biological roots—the conquerors are trying to spread their genes. This film is so objectionable that the sociobiologists who are interviewed in the film disavow any connection with it. Granting their displeasure with the final version, it should be noted that their statements in the film do not seem to have been taken out of context and are not at all at variance with the visual scenes shown while they are speaking. But, even supposing that they were unwitting instruments of the producer, can sociobiologists disclaim all responsibility for a film that simply extends standard sociobiological reasoning?

The protestations by sociobiologists that they intend no political or ethical implications in their theory are undercut not only by the film but also by the fact that sociobiologists have publicized and politicized their theory by means of articles and interviews in such popular publications as the *New York Times, House and Garden, Newsweek, Time,* and *Psychology Today.* It hardly needs to be mentioned that these publications have concentrated on human, rather than animal, sociobiology.

The consequences of sociobiology take two forms. First, the theory, like Jensen's, has implications for official decision making. For example, sociobiology might be considered relevant in deciding whether affirmative action programs for women should be enforced if women are biologically less suited for professional careers. Second, the theory affects the attitudes and prejudices of people as they lead their daily lives. Will they treat others as equals or as inferiors? We are well aware of the distinction between the theory and its vulgarizers, but where does responsibility begin or end? Using any sort of balancing test, the benefits accruing to human knowledge and to society gained by publicizing the highly speculative theory of sociobiology as scientific fact and by encour-

aging its use by social scientists are in our opinion clearly out-
weighed by the actual and potential harm resulting from the theory. It seems essential that sociobiologists and those who would use
sociobiological concepts be fully aware of the implications of their
theorizing. In addition, all of us must take on the responsibility of
becoming familiar with the theory, its flaws, its implications, and
the results of its use. The social impact of such doctrines as sexism
and racism is too great to be ignored.

11 *Marjorie Grene*

Sociobiology and the Human Mind

By request, I will address the "methodological problems of providing biological explanations for mental phenomena." Scientists in general, I am afraid, will suppose that philosophers professionally resist such explanation, claiming for themselves some "intuition" of a truth exempt from scientific testing and proclaiming, at least implicitly, the existence of some sort of thing called *mind* that is supposed to exist apart from matter. Since I do not myself subscribe to the view that philosophy provides infallible intuitive wisdom and since I cannot imagine what disembodied mind could possibly be, I am saddened by this prospect. At the same time, occasional conversations with sociobiologists or biologists sympathetic to their point of view have only confirmed my fear in this respect. It is in a spirit of hopelessness, therefore, that I shall nevertheless attempt to formulate some objections to sociobiological explanation of those human concerns and activities sometimes referred to as *mind*.

First let me specify a little more precisely the terms of our problem. By "biological" explanation in this context, I understand evolutionary explanation, and by evolutionary explanation, I understand explanation in terms of relative fitness—that is, of the probability that one gene or set of genes rather than another will persist in the next generation of a given population. The relative fitness of particular genes is what "biology" in this context is concerned with, and nothing else; not that that really *is* what biology is exclusively concerned with, but we are speaking here of *sociobiology*, which explicitly confines itself—and prides itself on doing so—to a strict, orthodox selectionist framework in its inquiries. Not that Darwinism need be so restricted, but that is how sociobiologists see it (S. J. Gould, 1977; Mayr, 1974).

And what are "mental phenomena"? They are, I presume, events or experiences or dispositions or activities usually associated with "mind" rather than "body," such as solving or being able to solve a mathematical problem, or composing a sonnet, or telling right from wrong, possibly also emotions or character traits such as ambition, hope, irascibility, or a sense of humor. And what is to be understood by the "mind" to which such activities or dispositions are ascribed? I shall adopt here, as I have elsewhere, a definition by A. J. P. Kenny, adding a few revisions to clarify, if I can, what I think is meant. "To have a mind," Kenny writes, "is to have the capacity to acquire the ability to operate with symbols, in such a way that it is one's own activity that makes them symbols and confers meaning upon them" (Kenny and others, 1973, p. 47). The word *symbols* here refers to artifactual entities, assigned by convention to assume certain roles in the social practices of their users; and the word *activity* here means responsible activity, characteristic of the kind of center of action that can do right or wrong, be praised or blamed. All this needs to be spelled out further, but I hope the intent is moderately clear. It should be emphasized, moreover, that such a capacity, although taken as definitive of mind, is itself psychophysical in the sense that it is (by definition) constitutive of the human psyche but at the same time the capacity of a certain kind of living body. On the one hand, mind is not a *thing* separable from the body whose capacity it is. Nor, on the other hand, is mind a secret inner something of the kind Wittgenstein accused philosophers of looking for (and so they did—but they need not have!). If I understand Donald Griffin's thesis about animal awareness, he is concerned with mind in a sense at least remotely related to this "secret inner something" (Griffin, 1976). And, of course, many other animals, too, feel pleasure, pain, frustration, and curiosity; one infers also that they make inferences of a sort, and so on. They live at the center of an *Umwelt* presumably experienced differently from ours but nevertheless experienced (Grene, 1974, chaps. 16–19). But it is precisely *not* mind as subjectivity that is at issue here; it is the human mind, which is characteristically founded on the responsible use of conventionally based symbols. (If Washoe and Sara turn out to be symbol users in this sense, fine—that may be an empirical question. Bees, it seems to me on the evidence, are much less likely candidates. Even on evolutionary principles, one supposes those massive primate brains are there for something [Geschwind, 1974].) Let us look very briefly at the sort of processes or dispositions one might describe as "mental" and in particular at the place of subjectivity in such allegations. We may hold, for example, as an ethical principle, that it is wrong to give pain unnecessarily; it is the entertainment and acceptance of the principle that is mental, rather than the pain, imagined or felt. True, the tooth*ache* I feel is the so-to-speak "mental" aspect of my decaying tooth. But this is mentality in a trivial and evanescent sense. If one talks about the quality of a person's mind, one does not list his or her pleasures and pains or even long-term desires and emotions. One does speak, perhaps, of buoyancy or melancholy, vividness or lethargy, but primarily one speaks of quickness or slowness of thought, incisiveness or muddle-headedness—that is, of abilities related to the responsible use of humanly instituted symbol systems, whether in the sciences or the arts or in the worlds of practice or of theory, both of which are publicly and hence institutionally constituted—that is, artifactually and therefore symbolically. Space does not permit me to argue this thesis here (although the last point I shall make is related to it), but it is important to insist that "mind" need not—indeed, should not—be understood either as substance or as subjectivity: as a separate entity, a *res cogitans,* as Descartes defined it, nor yet as an aggregate of private feelings, as empiricists have tried to interpret it.

Our question, then, is "Are there methodological obstacles

to explaining mind in the sense proposed through reference to evolution in terms of relative fitness?" Hard-line sociobiologists see none; let me suggest half a dozen.

First, one of the powers mediated by the ability to initiate and rely on human symbol systems is the power to make statements—whether relatively factual statements or relatively theoretical ones—of which we claim, in making them, that they are credible: that, on the evidence, they ought to be believed. Sociobiology—"a new synthesis," as it vaunts itself to be—claims that on principle, when knowledge of the neuron and the gene have been perfected (in a hundred years—the time span is irrelevant), it will embrace within its scope all human activity. This will include the activity of scientists as scientists. Now, any theory that tries to account for cognitive claims, such as those made by sociobiology, must itself stand or fall as knowledge on its own grounds. Hume reduced most of what had counted as knowledge to the mere force of habit. And I believe he recognized his own argument in the *Treatise* (Hume, [1739] 1975) as the issue of the force of habit in his own case. But sociobiology provides no equipment for such self-reference; it claims a credibility that on its own grounds it allows to no statements. In other words, it entails an explanation of cognitive claims that makes its own cognitive claims impossible in the form in which it makes them. For if they carry through their own theory sociobiologists should recognize that their "arguments," which, like all behavior, are the effects of evolution in terms of relative fitness, are founded not on reasons for holding that their statements are credible but on a tendency toward maximization of reproductive efficiency on the part of some of their genes. But the same holds for their critics' statements. So there is no argument, only the universal war of gene against gene. This kind of theory, therefore, like every extreme reductivism, is epistemologically self-defeating.

Mind you, that is not—repeat, not—to say that scientific knowledge can have no bearing on philosophical reflection. The trouble with sociobiology is not so much that it is science with unwarranted philosophical pretensions as that it is—and this is my second point—that it is antiquated science. True, the fashion began with Hamilton's paper of 1964, and so it may seem relatively novel. But in its methodology it is Newtonian in at least four respects. (When I remarked on this a couple of years ago to one of Wilson's disciples, who had been lecturing to my class, he replied happily, "Yes, Newton is our god!") First, sociobiology naively accepts Newton's laws of causal reasoning. But these laws did not have to wait for Einstein and Planck to undermine them; the job had already been done by Berkeley in the *De Motu* (Berkeley, [1721] 1968) and by Hume in the *Treatise* (Hume, [1739] 1975). Rational causal explanation does not follow simply from regularity, nor do statements about some of the phenomena lead us with necessity to statements about all of them. Moreover, in addition to its naive acceptance of regularity as yielding explanation, Newton's naive atomism, too, is retained in sociobiological thinking: Genes replace hard, solid, impenetrable particles as ultimate units. And so indeed (third) is a very simple faith in the mathematical nature of reality. As Kepler's second law was provable in terms of the geometry of similar triangles, so our behavior is, according to sociobiology, demonstrable in terms of cost-benefit analysis. Yet mathematical applications need not always work as smoothly as they did for the cases with which Newton was dealing. A model, by its very economy, may fail to capture the phenomenon it is seeking to explain. (For example, see Richard Levins' excellent papers on the use of models in ecology [Levins, 1966, 1968, 1975].) Besides (fourth), even Newton's methodology, with its limited subject matter, was supported and rendered rational only by God's providence. Wilson's, one supposes, lacks such supernatural foundations.

Further (third), the new mechanical philosophy of the seventeenth century did not in itself presume to give an account of mind. Many people then and later tried to turn it to this use, but usually, like Hume, by analogy—"an attempt to introduce the experimental method of reasoning into moral subjects"—not often, like Hobbes, by a simple absorption of the human *into* the natural world. But just such an absorption is what Wilson, in his bestselling book, assumes.

Hume's *Treatise of Human Nature* ([1739] 1975) was subtitled *An Attempt to Introduce the Experimental Method of Reasoning into Moral Subjects*. It seemed to many—as *a fortiori* it seems latterly to sociobiologists—that the methods of the then "new mechanical philosophy" ought to prove as successful for the formulation of the laws of thought as they had proved, thanks to the incomparable Sir Isaac, to the laws of motion. Much of modern social science, of course, still rests happily on this assumption—even though some social scientists and many philosophers of social science have shown how, through the change of subject matter from material things or forces to human agents, even the methodological analogy is bound to break down. And David Hume, being as honest as he was amiable, had already confessed to the breakdown, for his atomizing epistemology was unable, he admitted in the appendix to his *Treatise,* to deal with the problem of the person.

Almost a century earlier, Hobbes had tried more dogmatically and less candidly not only to use the new (in his time, of course, Galilean, not yet Newtonian) method for the explication of natural phenomena but also to assimilate "moral" into "natural" philosophy, to explain mind *as* matter in motion. It was a bold attempt and most unpopular, since like sociobiology it aimed at the reduction of all standards, purposes, or duties—all to which we owe allegiance—to the blind bombardment of particles by one another. Yet it is a vision that recurs when scientists or others, dazzled by science, see the movement of least bits of stuff as an enlightening and even liberating model of what there is. It collapses, of course, on the ground of my first objection, as well as on the ground of the objections to Newtonian method suggested in my second objection and, most basically, to its radical inability to deal with human discourse and human agency (my sixth objection, later). As Cynthia Schuster argued recently (1977), when Hobbesian motions produce promises and so institute "obligations" they create new entities that radically transcend their frame of reference. The Hobbesian enterprise is no better able than the Humean to bring human action—and with it, mentality—into the scientific fold.

Once more, that is not to say that the advance of science and especially of biology is irrelevant to philosophical reflection about the human person or the human mind but rather that the dicta of science should be thought about with care, not swallowed whole.

Fourth, what biology can give us to assist in philosophical

reflection about the nature of mind is mainly an account of the necessary conditions for its existence, as well as of the material units it manipulates, which are necessary conditions in a more limited although more interesting sense (Sahlins, 1976b, p. 66). We are indeed animals, and if we want to consider what strange sort of animals we are we should be aware of the limits within which our peculiar symbol-related abilities could have arisen and can be maintained. The study of the brain, for example, is not as such the study of mind, but the study of brain-damaged patients has certainly shed light on philosophical problems about mind, because it shows up, *a contrario*, the character of the "normal" conditions within which symbol-using ability is able to develop. Sociobiology, however, purports to answer philosophical questions about mind by specifying necessary biological conditions for its evolution. For example, we are supposed to have a theory of ethics when we know that our emotions (which may or may not be the source of moral judgments) developed in relation to the hypothalamus. This is to confuse necessary with sufficient conditions. If there are to be moral judgments, there must be animals to make them—and *perhaps*, although this by no means follows, those judgments are made possible not so much by the development of the "higher" as by that of the "lower" brain centers. I suspect this kind of division is crude even biologically; philosophically, it is on a par with the project of Jefferson's friend who wanted to dissect cadavers in order to locate the seat of the moral sense. To discover the biological conditions for mental development is not to say how, within those conditions, mind works.

By the same token, there is also a confusion here between the empirical and the conceptual, for not even a description of a set of sufficient empirical conditions would provide on its own a conceptual analysis of mind. Philosophical problems are always metaproblems, reflective questions about the significance of some kind of entity or activity or of its conceptual structure under some description. The moral philosopher wants to know what it means to make moral judgments, not how the creatures making such judgments came to be. Evolutionary discoveries cannot answer questions of ethical theory; they are simply tangential to them. Again, this does not mean that advances in empirical knowledge can have no bearing on philosophical problems, but that bearing is always indirect and always partial, for the questions and answers in the two cases are not on the same logical plane. This is a point that sociobiology entirely fails to grasp.

Fifth, even where biology does bear on philosophical problems, it need not by any means always be evolutionary biology that is in question. Neurophysiology is a case in point. Of course, the brain *did* evolve, but its study entails much more than the concepts permitted by evolutionary theory. At every stage in evolution except the very first, there are always already complex organized living systems in existence. Selection will decide which of these will last. But any of these, once they exist, may be studied in relation to questions other than the question, how they arose. One may ask of such a system, for example, what it is made of, or how it works, in whole or in part. (Incidentally, the study of neurology is not only, as Wilson seems to think, the study of neurons but is also the study of their organization.) In many of its practitioners, however, a preoccupation with evolutionary explanation seems to act like blinders on a cart horse: It focuses attention on the road and conceals frightening movements in the ditch or in the fields, but it also conceals much that a less blinkered vision might gain from looking at. Why should the explanation of a mechanism in terms of its origin always be the "ultimate" explanation, while, say, a physiological account of its operation would be only "proximate"? One might, with a rough guess at its mode of functioning, make an inference about its evolutionary beginnings and then get down to work on a detailed study of its operation. Even within biology, then, evolutionary explanation provides necessary conditions for the existence of a phenomenon, not an understanding of the phenomenon itself—the latter would constitute, relatively speaking, an ultimate explanation (although there is, of course, no *ultimate* explanation!).

Sixth and finally, the gravest error of a sociobiological approach to the problem of mind lies in its failure to understand the nature of artifactual symbol systems and of human beings as animals dependent for their peculiar life-style on the existence of such systems: One could say, for short, its failure to understand the way in which mind is an expression of culture. In this respect, Marshall Sahlins' *Use and Abuse of Biology* (1976b), especially his second chap-

ter, goes in my view to the heart of the matter, and what I have
to say here is chiefly a recapitulation of part of his argument. Sahlins is defending culture against a reduction to nature; the two are
hierarchically related, as biology in turn is to physics. I have been
arguing against a reduction of the mental to the biological, in particular to terms of evolutionary biology. What makes culture unique,
however, in Sahlins' view, is its "construction by symbolic means"—
that is, by arbitrary means—in which some things have been made
to stand for others, not necessarily or biologically, but in any of an
indefinite number of possible ways. That is precisely the defining
feature of mind on the view I am advocating. True, culture or
human society and the mind of the individual are not identical;
but the mind of any given individual is one expression, however
unique or original, of the symbolic structure of his or her society
as distinct from other societies constructed in the light of other
symbol systems. Thus Sahlins' argument against the sociobiological
reduction of culture constitutes at the same time an argument
against the sociobiological reading of mind.

Sahlins presents a series of empirical theses that combine to
suggest a philosophical conclusion. Firstly, "no system of human
kinship relations is organized in accord with the genetic coefficients
of relationship as known to sociobiologists" (Sahlins, 1976b, p. 57).
Secondly, "as the culturally constituted kinship relations govern the
real processes of cooperation in production, property, mutual aid,
and marital exchange, the human systems ordering reproductive
success have an entirely different calculus than that predicted by
kin selection" (p. 57). Thirdly, "kinship is a unique characteristic
of human societies, distinguishable precisely by its freedom from
natural relationships" (p. 58). Fourthly, "human beings do not reproduce as physical or biological beings but as social beings," and,
fifthly, "what is reproduced is not human beings *qua* human beings
but the system of social groups, categories, and relations in which they live"
(p. 60). In conclusion, then, "culture is the indispensable condition
of this system of human organization and reproduction," and culture in turn is defined in reference to its unique "construction by
symbolic means." Sahlins quotes Wilson's statement that "culture,
aside from its involvement with language, which is truly unique,
is different from animal tradition only in degree" (Sahlins, 1976b,

p. 60, quoting Wilson, 1975a, p. 168), and makes two comments on it. One, if we were to disregard language, the main clause of the statement would of course be true: Without language—*our* kind of language—we would be no different from other animals, just cleverer in some ways, less clever in others. But it is precisely the relation to language (and the power to construct arbitrary symbol systems that makes language possible) that differentiates culture from animal tradition in kind rather than degree. The same comment holds if we are concerned with the mental rather than the cultural, with mind as the individuation of culture. Two, where Wilson does deal wih language, he takes its importance to lie in its *"function of communication"* rather than in its *"structure of significance."* Speech is here understood as intended *"to convey information* rather than to *generate meaning"* (Sahlins, 1976b, p. 61). As communication, however, Sahlins points out:

> Language is not distinguishable from the class of animal signaling; it only adds (quantitatively) to the capacity to signal. What is signaled is information—which may be measured, as in classic theory, by the practical alteration in the behavior of the recipient from some otherwise probable course of action. . . . This functional view of language, which incidentally is exactly Malinowski's, is particularly appropriate to a biological standpoint, for by it human speech is automatically subsumed in the adaptive action of responding to the natural or given world. What is lost by it is the creative action of constructing a human world; that is, the sedimentation of meaningful values on "objective" differences according to local schemes of significance. So far as its concept or meaning is concerned, a word is not simply referable to external stimuli but first of all to its *own* environment of related words. By its contrast with these is constructed its own valuation of the object, and the totality of such valuations is a cultural constitution of "reality" [Sahlins, 1976b, p. 61].

It is difficult to overestimate the importance of this distinction. Insofar as language consists in conspecific signaling, our achievements are only quantitatively different from those of other animals. If we can transmit signals home from the moon, as we assume that bats or bees cannot, they can nevertheless signal in ways until recently undreamt of and that lie entirely beyond our powers. It would be folly to deny both the variety and scope of animal communication. Human language, however, does not consist entirely in signaling. Our discourse and the symbol systems it makes possible are more than an elaborate Morse code. Admittedly, it is hard indeed to say directly, let alone briefly, *what* the more is that they are. Although the point Sahlins is making is extremely important, his terminology is unfortunate. It is not encounter-groupish "meaningful relationships" or "values" (whatever they are), somehow superimposed on signaling devices, that make us human. The work of such symbolic anthropologists as Sahlins himself (Sahlins, 1976a) can exhibit for us something of the elaborate artifactualities, the coming to ourselves within artifacts, that—as far as we can tell—distinguish human practices from animal behavior. So can Heidegger's account of *Dasein's* everyday being (Heidegger, 1927), Merleau-Ponty's description of our embodied being-in-the-world (Merleau-Ponty, 1945), always both wholly cultural and wholly natural, or Helmuth Plessner's essays in philosophical anthropology, let alone his greatest work, *Die Stufen des Organischen und der Mensch* (Plessner, 1928, 1970). I have expounded some of these arguments elsewhere and need not repeat them here; let me just say again, with Sahlins, that the linguistic and language-borne structures within which, and in dependence on which, human life develops far outstrip the communicative needs of the species. As communicators, we are on a par with other animals again: cleverer in some ways, not so clever in others. It is as dwellers within the inexhaustible artificialities of natural languages—each different from all others and, in its shifting history, from itself—that we appear to be very peculiar animals indeed. My first point is that it is not as communication but as something else— as poetry, perhaps, a home built of the tenuous threads of metaphor—that human language needs to be understood.

Finally and most fundamentally, language as a web of metaphor is the expression of human historicity. Another brief passage from Sahlins may make this point: "What is here at stake is the understanding that each human group orders the objectivity of its experience, including the biological 'fact' of relatedness, and so makes of human perception and social organization a *historic* conception. Human communication is not a simple stimulus-response syndrome, bound thus to represent the material exigencies of survival. For the objectivity of objects is itself a cultural determination, generated by the assignment of a symbolic significance. to certain 'real' differences even as others are ignored. On the basis of this segmentation . . . the 'real' is systematically constituted, that is, in a given cultural mode" (Sahlins, 1976b, pp. 61–62—my italics). But it is, again, the capacity to acquire the ability to enter into just such symbol-constituting and symbol-constituted activities that is definitive of mind and so renders mind, like culture, irreducible to its biological and in particular to its genetic conditions.

It should be added, in conclusion, that Wilson himself, so far as I can understand his very murky argument, seems to have moved away (Wilson, 1977) from the more confident (and more dogmatic) theses of his original program (Wilson, 1975a). This is a welcome, if confusing, development, that confirms one's hope that this particular "new synthesis" may soon join the other interesting relics (described by J. B. Schneewind in this volume) that lie about the lumber room of our intellectual history. *Requiescat in pace.*

12

J. B. Schneewind

Sociobiology, Social Policy, and Nirvana

$\mathcal{T}$he urge to find a way of bringing all our knowledge of the universe into a system grounded on a single principle is as old as speculation. It is an understandable development of the scientific need to explain as much as possible by as little as possible. Kant argued that there is an inevitable tendency of human thought to extend empirically based theories and modes of reasoning to matters beyond the reach of experiment, and theoreticians do in fact display this tendency even more frequently than generals show a desire to try to conquer the world. E. O. Wilson's (1975a) program for sociobiology displays some of the characteristic marks of such a theoretician's ambition. He is, of course, not the sole proponent of sociobiology nor the only one to articulate imperial claims for it. He has, however, stated his views candidly and more fully than

I would like to thank professors Daniel Bates, Virginia Held, and Susan Lees for their helpful discussions of various aspects of this chapter.

many other exponents of the general position, so that it is worth examining his work to see if we can tell what these imperial claims might come to.

Wilson proposes to increase our understanding of the behavior not only of insects and lower animals but also of the higher animals and humans. He wants to include what he calls "the humanities," ethics, and religion, as well as the social sciences, in the "new synthesis" (Wilson, 1975a, pp. 3–4, 129, 560–61.) He remarks a number of times on the differences between humans and other animals, especially noting those due to the human use of language (Wilson, 1975a, pp. 202, 380, 550, 551, 559), but his awareness of these differences does not lead him to make any qualifications of principle in his program. To judge from some of his later statements in interviews and elsewhere, Wilson seems to be having second thoughts, and even in *Sociobiology: The New Synthesis* (1975a) itself the program is not laid out with the kind of precision and clarity a scientist would expect from a specific research proposal. It is therefore difficult and perhaps unfair to say positively just what Wilson expects the new synthesis to cover or how he expects the material to be included. Yet it is hard to avoid the conclusion that he believes at least that somehow it is possible to derive a rule of life for humans from the laws governing the behavior of termites and turkeys (Wilson, 1975a, pp. 129, 562–564). One need not question his genetics, his biology, or his zoology to raise objections to his apparent program as a whole. Several anthropologists have argued convincingly that there are difficulties both in principle and in detail about the claims of sociobiology concerning the understanding of human society. A certain amount of heat is evident in response to the reductionist claims Wilson makes. But what arouses the warmest emotions about Wilson's program and what seems to be responsible for the general public interest in it (Galvin, 1977, pp. 54–63) is not its claim to be able to *explain* human action. It is the thought that sociobiology can give *guidance* to human action—that it has implications for individual morality and for social policy. This is what gives point to the frequent comparisons of Wilson's work to that of Herbert Spencer. Despite the fact that Wilson has disavowed Spencerian intentions, his book sometimes seems—perhaps inconsistently—to display them. Spencer did try to show that certain moral rules and social policies applicable to specific issues in his own times had a basis or warrant in fundamental truths about evolution. And some of Wilson's critics have supposed him to be trying to provide a new foundation for similar kinds of thinking. It may therefore be helpful to look at Wilson's work in the light of earlier efforts to relate our knowledge of the world of plants and animals to our morality.

The great religious outlooks have always offered the kind of comprehensive view of nature, man, and society toward which Wilson seems to be striving, and a comparison with a philosophical articulation of a religious outlook is a good place to start. Richard Hooker, the last great English writer to expound the medieval world view, is a clear case. There are laws governing every kind of being in the universe, he holds, and these laws are all God's commands. They direct each kind of thing to some good operation, and the goods of each kind of thing fit in harmoniously with the goods of every other kind, so that the totality is a display of the glory of the Creator. Each part of creation ought to obey its proper laws. Stones and stars, termites and turkeys do obey, for the most part, although the existence of anomalies such as comets and two-headed calves shows that even the nonhuman parts do not always obey. So it is also with humans. God laid down the laws we ought to follow, and we do—mostly. We are unlike the lower creation in that we are consciously aware of God's laws and are able deliberately to refuse to follow them. Our knowledge of the moral laws is due to God's having given each of us the natural ability to know them, simply by consulting the "inner light" of our conscience. Like all natural laws, they direct us to our proper good. This good is peace or blessedness in union with God. It is a good in which all can share. There is no need to compete for it. If we are egoistic, if we try to keep for ourselves something we deny others, we do so either because of willful ignorance of our true goal or because of the weakness of will that is part of our imperfect nature. If we followed the laws of morality, we would find, in the end, that the good of each is inseparable from the good of all. The order of the moral life is simply an aspect of the order of the universe as a whole, and the laws of morality are no different in kind from the laws of the rest of creation (Hooker, [1594] 1845, Bk. 1).

Hooker summed up the tradition of a thousand years, but the foundations of his view began to become old-fashioned shortly after he published it in 1594. The emerging new science used a different concept of natural law from Hooker's, a concept that generated a deep problem for any systematic view such as his. It was the philosophers rather than the scientists themselves who worked out the problem, and of the philosophers it was, arguably, Spinoza who was most thorough. In his *Ethics,* published in 1677, the first aim was to show that the universe is a completely deterministic system, in which all parts, including humans, interact with all others according to unvarying laws. He explicitly rejected the idea that there is anything unique about humans, and he criticized those who "believe that man disturbs rather than follows [Nature's] order, that he has an absolute power over his own actions, and that he is altogether self-determined" (Spinoza [1677] 1949, Pt. 3, Pref.). He did not, like Hooker, think of laws as commands that are given for a purpose. He thought of laws in much the way modern scientists do, as the basic structure of the way in which things always do in fact behave. Any appearance of purpose or goal in nature is due to the human tendency to project our own wishes on our own imperfect understanding of the world. Spinoza's second aim was to draw the implications of his position for human action. It is striking that, although the work was entitled *Ethics* and although Spinoza was seeking wisdom for the guidance of life, nothing was said in it about what people ought to do. Indeed, the very concept of "ought" does not appear in it. Spinoza tried to show that anyone who has a full understanding of the way things are will see that one's own desires are not basically different from those of others and that there is no more or better reason to try to obtain the satisfaction of one's own desires than there is to strive to obtain the satisfaction of those of others. This understanding will alter the behavior of such a person, who will become what we think of as wise and who will find a new kind of peace in living. But Spinoza nowhere says that we ought to act as a wise man would. To the extent that we ourselves become wise, we will in fact act in that manner; to the extent that we fail to grasp wisdom, we will fail to do so; but Spinoza does not tell us to alter our behavior. In his political writings, Spinoza suggests various reasons why people have come to think that there is a point to telling others and themselves that they ought to behave in this way or that, and he suggests that there are situations in which it makes sense to tell people what to do. But the "ought," instead of indicating any final reality in the universe, results rather from the limitations of our ability to understand what there is. Spinoza, like Hooker, propounds a unitary system of law; but, unlike Hooker, Spinoza does away with anthropomorphic and teleological interpretations of law. All law is the kind of law determining inexorably what was, is, and will be. Morality as popularly understood is a result of our failure to grasp this.

Against this position, Kant asserted the fundamental, irreducible, and essential nature of the "ought" in the life activities of humans and of any other rational creatures there might be. The "ought" articulates a demand found by each of us in our own conscious experience. The demand is set by our own power of reasoning—the same powers that, in the realm of theory, set the demands embodied in logic and enable us to discover the causal laws governing inanimate objects. In practice, reason demands as a minimum that we limit action to satisfy egoistic desires in such a way that our own satisfaction would be compatible with the satisfaction of everyone equally. The common feeling that one ought not to make an exception for oneself exemplifies the way in which practical reason makes its demands felt in experience. But, although each of us is directly aware of the requirements of what we ought to do, moral theory must go beyond experience, Kant thought, to explain how such an "ought" can be valid for us. We must suppose a unique kind of freedom belonging to rational agents. The laws depending on this kind of freedom are different in kind from the laws of necessity that structure the inanimate world, and so Kant, unlike Hooker or Spinoza, denied the possibility of a unitary system of law and insisted that morality belongs to a realm of law different in kind from the laws governing nonrational nature. Because of this—because the demands of the moral "ought" are of a special kind—moral directives and principles cannot, according to Kant, ever be reduced to or derived from

the laws that govern the nonrational parts of the world. This claim
is one that every serious system builder since Kant's time has tried
to take into account.

Spencer certainly tried to deal with the claim. He deliberately set out to construct an all-encompassing system culminating
in ethics, because he felt one was needed. Like many other Victorian intellectuals, he thought that traditional Christian beliefs
were losing their hold on the mass of mankind, and it seemed appropriate to him to replace them with others that would have the
same scope and serve the same purposes. This was particularly
urgent for moral beliefs. In the preface to the first part of his systematic treatise on ethics, he said that "the establishment of rules
of right conduct on a scientific basis is a pressing need. Now that
moral injunctions are losing the authority given by their supposed
sacred origin, the secularization of morals is becoming imperative.
Few things can happen more disastrous," he added, "than the
death and decay of a regulative system no longer fit, before another and fitter regulative system has grown up to replace it"
(Spencer, 1896, xiv–xv). Spencer did not propose to sit idly by
while society evolved a proper replacement for its moribund morality. He proposed to generate one full-grown from his own head.
He proposed, moreover, to generate it out of a fundamental principle of evolution that governs all natural occurrences. He had little
to say about evolution as it applies to inorganic phenomena, but
he wrote voluminously on the involvement of the law of evolution
in biology, psychology, and sociology. The results of his studies of
these domains helped him develop the substance of his ethics. The
link between the descriptive studies and the normative ethics consists of two parts—a definition and a psychoevolutionary theory.

The definition forms the central step. The term *good,* which
Spencer took as basic, just means (roughly) "productive of an aggregate of desirable feelings" (Spencer, 1896, vol. 1, iii), and *ought*
can be defined in terms of productivity of what is good. Spencer
did not think these points needed much argument. He was more
interested in the way in which people come to have the beliefs they
have about what they ought to do, and here he constructed an important theory. He did not deny that we have the experiences of
ineluctable "oughts" of which Kant made so much, but instead of

seeking a transcendental account of them he offered an evolutionary explanation. His account was most clearly and succinctly given in a letter written in 1863. We do indeed have moral intuitions about what we ought to do, Spencer said, and these seem to us to enable us to grasp timeless moral laws. In fact, these "intuitions" are feelings, which are the results of thousands of years of accumulated human experience of the conditions that produce pleasant or painful experiences, that is, good or bad experiences. They now have a physiological basis, of which we are unaware. Hence in our present consciousness they do not appear to us as they really are: "Though these moral intuitions are the results of accumulated experiences of utility, gradually organized and inherited, they have come to be quite independent of conscious experience. . . . These experiences of utility, organized and consolidated through all past generations of the human race, have been producing corresponding nervous modifications, which, by continued transmission and accumulation, have become in us certain faculties of moral intuition—certain emotions responding to right and wrong conduct, which have no apparent basis in the individual experiences of utility" (Spencer, 1904, vol. 2, p. 89). Spencer transformed the individualistic psychology of the associationist school into a historical, species-wide psychology and thereby offered a more plausible account of the unique moral feelings that humans now have. And this, together with the definitions that reduced "good" and "ought" to concepts referring to pleasant experiences and their production, enabled him to argue that evolution could not only explain our present moral beliefs but could also give us guidance in constructing a new set of moral rules—a set that would now be shaped deliberately in accordance with a scientific understanding of evolution. The manner in which he thought this could be done is worth noting briefly.

Like most of his modern predecessors in ethics, Spencer took as central the problem of relating egoistic desires to socially acceptable behavior. But, unlike them, he did not see it as a permanent problem arising out of human nature. He saw it, rather, as representing a passing stage in the evolution of society. For evolutionary theory demonstrates, Spencer thought, that humanity is moving toward a condition in which each person will have only desires that are compatible with the satisfaction of the desires of others. Social conflict is on the evolutionary way out. When human conduct has evolved to what Spencer calls its "limit," people will be maximally useful to one another, abstaining from harming others and even positively cooperating with them, without the need there now is of sanctions to induce such behavior. Humans will be so programmed by evolutionary selection that the voluntary acts that satisfy the agent's own desires will also help satisfy the desires of other agents, and no one will have desires that require forwarding his or her own good at the expense of the good of others. Now, society as it will be at the "limit" of evolution provides us with a clear model of what a perfectly good society is, since it represents the arrangements through which desirable feeling will be maximized and undesirable feeling minimized. Hence we can use our scientifically based knowledge of the perfectly evolved society as a standard against which to criticize the presently accepted code of morals and as a goal we can consciously strive to reach. Or so, at least, Spencer originally hoped. After he actually had tried to work out the program and show the connection between the rules and policies he favored and the science he thought true, he had to admit that evolution did not provide as much guidance as he had hoped. His criticisms of Victorian morality were neither novel nor profound, and the directions in which he urged society to move were in essence the same as those he had advocated in print some years before he discovered the principle of evolution.

If Spencer's results, in ethics, were meager, both the method he thought should be used in obtaining them and the basis of that method were at least clearly laid out. Can we say as much for Wilson's sociobiology? There are some obvious similarities between Wilson's announced program and Spencer's position, but there are also some gaps in Wilson's presentation that Spencer had tried to fill in. Thus Wilson believes that morality is just a matter of emotions flooding our consciousness as a result of the action of the hypothalamic-limbic center, which is itself the outcome of evolution (Wilson, 1975a, pp. 4, 563, 575). This might serve as a reworking of Spencer's explanation of the apparent moral intuitions now widely shared among humans. But where Spencer offered a reductionist definition of a central moral concept Wilson offers none. We do not know whether he would adopt such a definition or not. Wilson's casual definition of perfect societies as "societies that lack conflict and possess the highest degrees of altruism and coordination" (Wilson, 1975a, p. 314) reminds us of Spencer's view of what society will be like at the "limit" of evolution, but Wilson does not explicitly use the concept of a perfect society as Spencer uses it. Wilson's hint about creating a "genetically accurate and hence completely fair code of ethics" (Wilson, 1975a, p. 575), like his remark that "innate moral pluralism" makes "the requirement for an evolutionary approach to ethics . . . self-evident" (Wilson, 1975a, p. 564), indicates a desire to do for ethics something like what Spencer tried to do. But in Wilson's writing it is unclear what method is to be used to expand these intimations. We may not like the method Spencer uses for transposing the facts and theories of evolution into moral principles and social policies, but at least it is fairly clear. It is, for example, quite clear that his method does not involve simply deriving principles and policies from evolutionary facts by enshrining the latter as norms. Thus Spencer points out that "the brutal treatment of women has been universal and constant" throughout the evolution of civilization (Spencer, 1896, vi, chap. 1, sec. 428) but this emphatically does not seem to him to imply that brutal treatment of women ought to continue. No doubt Wilson would agree that it does not follow from the fact that male dominance is prevalent in most animal species that male dominance ought to be the rule in human societies. But then it is quite unclear just what use Wilson would make of such a fact in constructing a "genetically accurate . . . code of ethics."

The basic aim of sociobiology seems to be to generate a comprehensive explanatory theory of animal behavior. This is quite different from the aim of providing a basis for a normative or directive view that is meant to guide human action, and it may be that Wilson's chief interest in morality is an interest in explaining it, rather than in improving it. The construction of such an explanation seems to be the point of his frequent discussions of what he calls "egoism" and "altruism." But here again there is a difference of some importance between Wilson and Spencer. And it is not only Spencer with whom Wilson differs but also with all of us (sociobiologists perhaps excepted) when we talk of egoism or selfishness in contrast to altruism or generosity.

When Spencer deals with this issue and tries to show what its evolutionary resolution will be, he is concerned, as social and moral philosophers have always been, with benefits or enjoyments or goods that one individual, conscious agent can possess only if some other conscious agent does not and with the conflict that arises because there is a scarcity of such goods in relation to the demand for them. To be egoistic is to be selfish—deliberately to seek or keep for oneself alone what someone else needs or wants, knowing there is not enough for both. To be altruistic is to be generous—deliberately to give up to others what one needs or wants for oneself. Our morality tells us to be generous or at the very least to refrain from selfishness, yet our basic psychology (so it has been argued since at least the time of Hobbes) leads us to very great or possibly complete selfishness. How can human morality contain such a perplexing norm? And how can we be expected to live up to it? Spencer—however misguidedly—thought, as I have indicated, that evolution would provide a solution to the problem by altering human desires. The process of evolution, he thought, would select and perpetuate human agents who simply would not have desires and needs that lead to selfish actions. Although he sometimes talks of evolutionary selection as resulting in agents whose behavior tends to increase the sheer "quantity of life," regardless of how those agents enjoy that life, he is basically concerned with the "quality" of the life preserved—with its being generally pleasant to the conscious agents, whose desires will increasingly come to be such that they can be gratified without frustration of the desires of others. This is disarmingly optimistic, although it is hardly convincing, but it is at least an effort to deal with egoism and altruism as they really concern all of us and as they are important in moral theory and practice. Wilson, unlike Spencer, seems to be concerned neither with the consciously felt quality of life nor with egoism and altruism as we normally understand them. The point is worth elaborating.

We must first note the way the terms *egoism* and *altruism* are used in sociobiology. The sociobiologist starts with genetics in its modern version. To put it simply, biology now tells us that any one gene tends to cause the existence of other genes having the same structure as it itself has. This is to be taken as a law of the operation of the complex molecules we call *genes*. Now, the larger organisms within which genes occur sometimes behave so as to increase the probability that the genes they contain will in fact cause copies of themselves to exist, and sometimes the larger organisms behave so as to increase the probability that copies of the genes they contain will be caused by different genes, identical in structure to the genes they contain but contained in other organisms. The former sort of behavior is labeled *egoistic* and the latter sort *altruistic* (Wilson, 1975a, p. 117). These concepts are then taken to be serviceable for the explanation of selfish and unselfish behavior in humans. In particular, they are said to explain evolutionary pressures toward altruistic behavior, which Wilson assumes is the core of morality.

There are two important respects in which Wilson's use of the terms *egoistic* and *altruistic* is idiosyncratic. As Wilson defines them, they have no relation to the conscious desires or satisfactions of agents or to their knowledge of the consequences of their actions for others. But egoism and altruism as they normally concern us involve both of these. An effort to get something I did not want or think I would enjoy or find useful would not be considered egoistic, and if I were completely unaware that by trying to get something I wanted I was depriving someone else of it I might be accused of thoughtlessness, or stupidity, or of acting simply out of habit or reflex—but not of selfishness. Similar points are true of altruism. Moreover, in their sociobiological use the labels *egoism* and *altruism* refer only to ways of affecting the gene pool. Acts that have no effect on the gene pool must therefore be neither selfish nor generous. But this does not correspond to our usual way of using these terms, according to which all sorts of acts that have no bearing on the gene pool can be either egoistic or altruistic.

Perhaps we should conclude that Wilson is using *egoism* and *altruism* as metaphors. But, if so, it is hard to see why they are appropriate metaphors, even within genetic theory. If we must use such anthropomorphic terms, we might as well say that the gene is a compulsive or even monomaniacal self-replicator. Why should we call it *egoistic*? After all, it is not the individual gene that is preserved or enhanced by causing structural duplicates of itself. For all we know, the gene gets no fun out of creating replicas of itself to go spinning down the ages in other organisms. If we use metaphors attributing desires and satisfactions to genes, perhaps we should feel sorry for them. Perhaps, like compulsive nail biters or cigarette smokers, they wish they could kick the habit. Perhaps they would like to quit making copies of themselves and enjoy life while it lasts.

Once we allow the tempting metaphors to be used, it is easy to see how Wilson might suppose he finds a structural analog between the laws of genetics and the laws of morality. He must use a certain interpretation of morality, which takes egoistic considerations as central and derives other-regarding considerations from them. This interpretation was worked out by Hobbes, and used in less sophisticated forms by a host of eighteenth-century writers such as Helvétius and d'Holbach. Its basic principle is that human motivation is really always egoistic or selfish and that unselfish acts are beyond the range of human possibility. Consequently, the rules and laws of morality that seem to require something other than self-aggrandizement or self-preservation are in reality only instructions to go about preserving or increasing one's own good by indirect routes. It is agreed that there are widely accepted moral imperatives demanding that one not break promises, ignore laws, or take more than one's fair share, even if self-sacrifice seems to be required. The theory is that what these imperatives represent is not, in the last analysis, a demand for permanent self-sacrifice. They outline a shrewd policy of long-range investment in a stable society, which ultimately pays off for each of us better than do overt smash-and-grab tactics. There are several notorious difficulties about moral theories of this sort, but I do not wish to insist on them here. The points to note are, first, that such theories deny the existence of real altruism and, second, that they require the existence of really egoistic motivation in individuals. Now, if genes were, or caused their containing organisms to be, "egoistic" in any ordinary sense and if such a theory of morality were acceptable, there would be an interesting parallel between the laws of genetics and the laws of ethics. Direct replication strategies would correspond to crude egoistic policies, and indirect replication strategies would correspond to apparently altruistic moral dictates. But the denial of real altruism is questionable, and, more importantly, gene behavior is not in any sense really selfish or egoistic. Hence there is no parallel between the laws of genetics and the laws of morality and no reason to suppose that the former provide an account of the latter. Once we see genetic behavior for what it is—a mindless, limitless, compulsive drive toward making copies of oneself—we will not be tempted to suppose that it somehow parallels even a crude human morality. After all, limitless production of offspring regardless of circumstances has never been demanded by any human code. The whole appeal of Wilson's suggested explanation of ethics seems to stem from metaphors that in the end must be admitted to be question begging.

Wilson has not, then, provided any reason to suppose either that sociobiology can provide a basis for rational criticism and improvement of existing moral beliefs or that it can generate a scientific explanation of those beliefs. The vociferous and vehement critics who denounce sociobiology because it leads to pernicious ethical principles are as much in error as are advocates who suppose that it will yield a new and improved morality on a sound footing. Sociobiology as it has so far been presented has no more logical or rational implications for morality than did nineteenth-century versions of evolutionary theory. But the lack of such implications did not prevent thinkers in the past from supposing that evolution did have a bearing on morality, and it may not prevent present-day thinkers from following them. A new systematization of major diverse fields exerts a power over the imagination that is out of all proportion to its logical power and that is not easily checked by the kinds of rational considerations I have been advancing. More often than not, the power of such a vision depends on the ambiguous or metaphorical use of a key concept. To point out the ambiguity, as I have tried to do with Wilson's key terms, is not to criticize the scientific content of the system. It is only to point to its limits. Such criticism may not, of course, check the imagination of the enthusiast for the system, and probably there is little one can do to argue against the power of a vision. But one can at least try to put it in proper perspective, and I shall conclude by suggesting one important perspective on the vision underlying the ethical pretensions of sociobiology.

Wilson is in an important way correct in thinking that he is not proposing a new version of Spencer's doctrine. As I have indicated, Spencer was, after all, concerned about individual conscious humans. Wilson is not. Happiness and unhappiness as we feel them do not matter to him: only the makeup of the gene pool does (Wilson, 1975a, p. 4). For Wilson, individual consciousness is at best a secondary matter. This is by no means a novel view. If we go back beyond Spencer to Schopenhauer, to Spinoza, and to the wise men of India, we see something like a common tradition that stresses just this point. But there is a crucial difference between the imaginative vision underlying this tradition and that underlying Wilson's position. Schopenhauer, Spinoza, and the teachers of the East each, in different ways, sought to provide a means whereby a person could come to know the illusory nature of conscious individuality and the relative unimportance of private happiness or unhappiness and could thereby reach an accommodation with the underlying impersonal reality. This accommodation would result in the cessation of troublesome striving, in a lasting peace—or, at the extreme, in nothingness, self-annihilation, Nirvana. Such an outcome is itself intelligible as a goal of human striving, and it is just this that differentiates these views from the position suggested by Wilson's concern with the gene pool. To find peace, to be free of the unhappiness that comes from unsatisfied desires, to shed the entire burden of consciousness—this is a goal one can understand, and so one can feel, even if one resists, the power of visions of the world that hold forth such a promise. The metaphors of sociobiology, however, do not seem to offer even this much hope. As we follow Wilson's argument, we see ourselves forever captive inside a consciousness that is a by-product—that is, indeed, a tool— of the forces governing genetic, and therefore ultimately molecular, behavior. The entire upshot of these forces is simply to produce more molecules with certain structures—to perpetuate some kinds of nonconscious, suborganic entities—rather than other kinds. And this is not a state of affairs that is intelligible as a goal of human striving.

Perhaps it is this aspect about Wilson's vision that accounts for the pessimistic tone of his cryptic final paragraph. Spinoza thought that perfect knowledge of the deterministic nature of the universe would bring us peace. Wilson (1975a, p. 575), in striking contrast, expresses the fear that "when we have progressed enough to explain ourselves in . . . mechanistic terms . . . the result might be hard to accept." He thinks we may have another hundred years before we find ourselves in what Wilson, quoting Camus, calls "a universe divested of illusions and lights." If "the humanities," ethics, and religion are simply illusions and lights that will be stripped away by the advance of sociobiological science, this is a bleak vision indeed. It is as far removed from the austere hope held forth by a Schopenhauer or a Spinoza as it is from the facile optimism of a Spencer. It is a vision for which none of the science in Wilson's book provides any rational justification. It hardly seems a vision that can nourish the imagination.

13

Donald R. Griffin

Humanistic Aspects of Ethology

One of the stated objectives of this volume is to explore the reciprocal influence of science and humanism. In the context of the current discussion of sociobiology, I should like to examine the possibility of a convergence of ethology and some aspects of humanistic studies. We are accustomed to two distinct meanings or implications of the term *humanistic:* (1) the literal one—pertaining to our species—and (2) the meaning that is customarily contrasted to such terms as *materialistic, technological,* and *reductionist.* This second usage implies an emphasis on value judgments, esthetics, symbolic communication, and related subjective and mental states.

It is almost universally taken for granted that these two

Many scientists and scholars have helped me to clarify the views expressed in this chapter. Without implying that any of them agrees with my views, I should like to express my gratitude to the following for their constructive and stimulating comments: Donald T. Campbell and Randolph Blake ("The Question of Animal Awareness," *American Scientist,* 1977, *65,* 10); and Jonathan Bennett, Margorie Grene, Joseph Margolis, Gareth B. Matthews, Harvey Sarles, and John R. Searle.

meanings are congruent and that humanistic attributes are unique to *Homo sapiens*. But this assumption requires reexamination in the light of recent ethological discoveries about animal behavior that have reopened serious but long neglected questions about the common view that thinking is a unique human capability. Although it is obvious that our species engages in enormously more complex, versatile, and effective manipulations of symbols and ideas than any other form of life known to us, the general acceptance of evolutionary continuity means that a comparative analysis of human and animal thinking is scientifically appropriate, provided (1) that the latter does occur and (2) that it can be rendered accessible to scientific scrutiny.

Such terms as *thinking, awareness, mental experience,* and the like are necessarily vague and imprecise, primarily, I suspect, because we know too little about their physiological bases. I will begin by using these words in their commonly accepted meanings—including the breadth and uncertainty of those meanings—and attempt to sharpen and clarify meanings that seem appropriate for discussing the mental experiences of species other than our own. Four distinguished scholars and scientists have recently devoted twenty thoughtful Gifford lectures and discussion to "The Nature of Mind" and "The Development of Mind" (Kenny and others, 1972, 1973). But they refrained from formulating any rigid definitions of mind or consciousness. Their reasons were the variety of attributes and activities we refer to as *mental* and the practical difficulties of specifying their precise nature in scientific terms. Such attributes as intelligence, rationality, creativity, intention, choice, discrimination, selection of goals, and self-awareness were frequently cited as significant mental states or experiences.

In the first series of Gifford Lectures, the physicist Longuet-Higgins offered a partial definition of mind in the following terms (Kenny and others, 1972, p. 136): "The idea of a goal is an integral part of the concept of mind and so is the idea of 'intention.' An organism which can have intentions I think is one which could be said to possess a mind. And intention demands, it seems to me, more than just having a goal, because you might say bacteria had, in some sense, a goal—I mean, to grow, and divide, and so forth. But I think the concept of intention goes beyond this and involves the idea of the ability to form a plan and make a decision—to adopt the plan. The idea of forming a plan, in turn, requires the idea of forming an internal model of the world." In the second series, Kenny advanced a quite different type of quasi-definition: "To have a mind is to have the capacity to acquire the ability to operate with symbols in such a way that it is one's own activity that makes them symbols and confers meaning on them" (Kenny and others, 1973, p. 47).

Many animals carry out sequences of behavior that appear to involve intentions and plans, although decisive evidence is difficult to obtain. Predators often stalk prey, and sometimes, as among wolves and porpoises, the activities of two or more animals appear to be well coordinated. Tool using has been repeatedly observed in a variety of birds and mammals (Beck, 1975). Herring gulls in some areas have fairly consistent habits of carrying shellfish aloft, dropping them over hard surfaces such as rocks and roadways, and then eating the previously inaccessible soft parts. While this process is not perfectly efficient—prey is sometimes dropped on soft surfaces or left uneaten—these birds may have some notion of the likely results of this rather specialized behavior. Many animals dig burrows or construct various types of nests and shelters. Zoologists have gone to great pains in recent decades to avoid any suggestions that animals have plans or intentions, but the learning abilities of many mammals, birds, and even fishes are adequate to justify a reasonable expectation that when such animals have previously constructed burrows or nests they have some memory of both the process and the end result. Indeed, the contrary conclusion requires postulating that their learning abilities demonstrated in other situations are somehow laid aside during these important activities.

Monkeys and apes have been observed to make efforts that seem designed to secure food without attracting the notice of more dominant members of their group who would otherwise be very likely to secure the food themselves. Rüppell (1969) describes cases in which a mother arctic fox was competing for food with her several half-grown young; the latter resorted to rather drastic tactics, such as urinating in their mother's face, to obtain morsels. After such encounters, the mother on several occasions gave warning calls, otherwise used to signal dangers of various kinds, and when the young ran off secured the food herself. It is difficult to interpret such behavior without postulating at least short-term intentions and plans on the part of both mother and young as they utilized, in competition for especially tasty food, these behavior patterns otherwise employed in such different situations.

Turning to Kenny's definition of mind as the ability to confer meaning on symbols by one's own activities in using them as symbols, it is appropriate to inquire whether any animals employ communication signals that are specialized for this purpose and distinctly different from behavior exhibited under any other circumstances. If such symbolic communication signals do occur, are they part of a regular system of sharing information with fellow members of a mutually interdependent group? Whether the information is conveyed by sounds, visible displays, or tactile contact, it would help satisfy the spirit of Kenny's definition if a codified series of signals were used not only in one particular situation but also in a variety of contexts where it was mutually advantageous for cooperating members of a group to exchange information about what needed to be done to further their common interests. We know that animals do not literally talk, but do they ever use other kinds of symbolic communication systems satisfying these criteria?

Forty years ago, the answer would have been a resounding and confident no. But in at least one well-documented case, a system of symbolic gestures is indeed utilized by members of a social group of interdependent animals to inform each other of the location of foods, water, or materials used to construct shelters. The desirability of these materials under the conditions prevailing at the time is also communicated. Under special circumstances that arise so rarely that they have never before been experienced by the individual animals concerned, a radically new need may arise, when something never before needed becomes of overriding importance to the welfare—indeed, the survival—of the animals. At such times, the symbolic communication system is utilized in a completely new situation to inform fellow group members of the location and quality of something these individual animals have

never needed or communicated about in all their previous
experience.

What animals really employ a communication system of the
type just described? Many will suspect that the obvious answer
seems to be chimpanzees taught elements of human sign language
or perhaps some species of monkey in which interdependent societies are well developed. Or perhaps I have been describing new
discoveries concerning the cooperative hunting in packs as practiced by dogs and wolves. Alternatively, one might wonder whether
some ethologist has discovered new levels of cooperation and
acoustic communication in one of the more specialized groups of
birds, such as songbirds or crows. In fact, however, I have described in outline form the communication system of a species of
social insect, the honeybee. This statement will strike many readers
as surprising or even ridiculous. How could an insect with a central
nervous system weighing no more than 2 milligrams be supposed
to engage in such versatile and symbolic communication behavior?

The essential facts are the following: Beekeepers have known
for many centuries that worker honeybees that are bringing either
nectar or pollen back to the hive often engage in vigorous patterns
of rapid crawling motions that have come to be called *dances*. Karl
von Frisch (1967) discovered about thirty years ago that the geo-
metrical patterns of these dances are clearly correlated with the
distance, direction, and what can loosely be called the *desirability* of
the food source from which the forager has just returned. The
dances do not occur at all, or only to a very limited degree, unless
the colony is seriously in need of something. Ordinarily, this need
is a shortage of either carbohydrate or protein food. The most significant pattern of dancing is the figure eight-shaped *Schwanzeltanz*
or waggle dance, which is used only when the food source is more
than roughly 50 meters from the hive (the distance at which waggle
dances begin varies among genetic strains of honeybees). During
the straight portion of the figure eight-shaped pattern—between
circling motions, which ordinarily alternate first to the right and
then to the left—the bee moves her abdomen laterally at a rate of
thirteen to fifteen cycles per second. Other worker bees cluster
around a dancer and are vigorously stimulated by tactile contact
with the moving abdomen.

At intervals between dancing, food is regurgitated by a dancer returning from a good supply of nectar, and there is abundant opportunity for the odors of flowers to be transmitted from dancer to followers. The duration of the waggle runs varies from one to many seconds, and it is inversely proportional to the distance from hive to food source. The direction is indicated relative to the sun's position in the sky. Almost all the dancing is carried out inside a dark hive on a vertical surface of honeycomb. The direction toward the sun is indicated by waggle runs directed straight upward, away from the sun by straight down, and by appropriately intermediate directions if the food is to the right or left of the sun's position. These and many other details about the waggle dances and other patterns of communication used by honeybees are described in detail by von Frisch (1967, 1971) and by Lindauer (1971). Not only do the waggle dances indicate the direction and distance at which a food source is located (accuracy about ± 5 degrees in direction and ± 10 percent in distance) but the vigor and intensity of the dancing and the amount of sound that accompanies it are also correlated with the desirability of the food under the circumstances prevailing at the time. Thus concentrated sugar solutions elicit more vigorous dancing if the colony is in need of food, but when the hive is overheated the most urgent need is for water, which is used to cool the air by evaporation. Foragers now bring water into the hive and regurgitate small droplets, and other bees engage in a behavior pattern known as *fanning,* in which they circulate air over the droplets in a manner that serves to increase the rate of evaporation and lower the temperature. Under these conditions, the most vigorous dances describe the location of water sources.

Not only do the properties of the waggle dances correlate with the distance and direction of a food source but some of the bees that cluster around the dancer and follow her through parts of the dance pattern later also fly out to the approximate location of the food. Von Frisch's original evidence that information was obtained by bees attending dances has recently been greatly strengthened by the ingenious experiments of J. L. Gould (1975, 1976), in which he experimentally altered the direction indicated by the waggle dances. Recruits flew to test feeders in the direction indicated by the altered dances even when the dancer had actually returned from a wholly different direction. While the system may not operate with maximum speed or efficiency, there is little doubt that information is transferred from one bee to another and utilized by the recipient to search in at least approximately the right area.

The standard reaction to the facts I have just outlined is that, while the dance communication of honeybees is remarkably complex for insects, it is nevertheless a genetically programmed, "hard-wired" behavior pattern consisting of a limited repertoire of responses elicited by specific stimuli. This is how we have been accustomed in the twentieth century to view the behavior of invertebrate animals; they are not supposed to do anything as complex as using a symbolic communication system.

Lindauer (1955, 1971) added an important dimension to our understanding of this versatile communication system. Worker honeybees live only a few weeks, but at intervals of months or even years an entirely new situation may arise when thousands of worker bees move out of their former hive, along with a queen, to form a swarm. Workers that had previously gathered food now search instead for cavities suitable for the establishment of a new colony. On finding suitable cavities, they carry out waggle dances, not inside the hive but while crawling over the mass of bees constituting the swarm. The same code for distance, direction, and desirability is employed, except that the actual direction toward the cavity is usually indicated by a bee dancing on a horizontal surface, rather than being transferred to a direction relative to gravity. Lindauer also observed that individual workers that have found and danced about different cavities exchange information and sometimes change their behavior accordingly. For example, a bee that has visited and danced about a rather mediocre cavity may be stimulated by the more vigorous dances of one of her sisters. The first bee may then visit the cavity about which she has received information by following these dances. She then returns to the swarm and dances about the superior cavity. Thus individual honeybees can act both as transmitters and receivers of information, and their symbolic communication is influenced both by external stimulation and by messages from the dances of other bees. The end result of this process is that all or nearly all the dancers in a swarm come to concentrate their attention on one cavity. When this has been going on for several hours, the swarm flies off together to this new hive site. The versatile use of the waggle dances by the bees certainly "makes them symbols and confers meaning on them."

My point in describing the bee dances as I have done is to emphasize that had such symbolic communicative behavior been observed in an animal closely related to our own species we would be quite ready to interpret it as intentional communication. It is because of our reluctance to credit insects with such behavioral capabilities that we are reluctant to recognize the bee dances as anything approximating human language (for example, see Bennett, 1964, 1976). Of course, the known "vocabulary" of dancing bees is very small, although J. L. Gould (1976) has pointed out that in terms of information-transmitting capacity it is second only to human language. But do we really wish to make size of vocabulary a crucial distinction between animal communication and human language? Do we mean that any organism that has a vocabulary of N words qualifies as human, whereas if it attains only $N - 1$ words it is subhuman? This reduction to the absurd serves to emphasize that we expect something other than vocabulary size to be decisive. Many have suggested that grammatical structure, creativity, productivity, openness, or other attributes of human language are not found in the communication system of other animals. In other words, new messages or combinations of elements are said never to occur. Yet when a honeybee first reports about a new source of food or some other needed commodity she is combining the elements of the dance code in a novel way to create a new pattern of communicative gesture she has not used before. More striking is the application of the dance communication system to the situation faced by a swarm of bees. These animals are communicating about an entirely new set of requirements that are unprecedented in their individual experience.

Without attempting to minimize the enormous quantitative gap in vocabulary size, and combinatorial ingenuity, that separates the language of even relatively simple-minded persons from the most complex animal communication systems, it does seem clear that most of the qualitative elements of symbolic communication are present in some communicative behavior of birds, mammals, and especially in the symbolic dances of honeybees.

An important sociobiological consideration was recently suggested to me in an informal conversation with Jack Bradbury and others, namely that complex symbolic communication systems are advantageous only in species where social groups have attained a high degree of interdependence and division of labor. Only in such animals, it may be argued, is there enough advantage to using a versatile communication system to have brought about its evolution. Not only honeybees but also many of the ants communicate about food supplies and dangers facing the colony (Wilson, 1971; Hölldobler, 1974; Möglich, Maschwitz, and Hölldobler, 1974). Much of this communication is based on chemical signals, which are much more difficult to analyze in detail than the honeybee dances, but gestures and tactile stimulation are also employed in some cases, as when one ant grasps another and leads it to a food source. Social insects and human beings, while very remotely related in terms of their evolutionary histories, both satisfy this criterion. However, some people live in relatively small and simply organized social groups, and it is widely postulated that at the time when language first developed our ancestors also lived in groups no larger and perhaps no more interdependent than baboon troops or packs of Cape hunting dogs.

The Potential Importance of Animal Thinking

If other species have subjective feelings and awareness of relationships, these may be very simple and limited in scope, but nevertheless significant in many ways, both to the animals concerned and to us. Their importance for our species includes not only a more realistic appreciation of our relationship to the rest of the universe but also the possibility of analyzing whatever biological roots may underlie human cognition. This possibility may be testable through the comparative approach that has been so fruitful in evolutionary biology, ethology, anthropology, and related disciplines. To the extent that this approach proves to have any validity, it will have broad implications not only for biology in gen-

eral—and for zoology, ethology, and sociobiology in particular—
but also for the very definition of humanity.

The behavioristic taboo has forced students of animal behavior into a cul-de-sac where even the most cautious and tentative
approaches to animal thinking have been inhibited by the argument that, no matter what sorts of data might be gathered, a behavioristic interpretation will remain just as plausible as a cognitive
one. To maintain this position consistently, the behaviorists have
been obliged to reject human mental experiences as scientifically
meaningless. But the weakness of this position is self-evident to
anyone from even the most rudimentary introspection, and many
humanists have recognized the crippling limitations of behaviorism—for example, Mackenzie (1977). Skinner (1974, p. 186) recognizes the existence of human mental experiences but seems to
reject the possibility that anything comparable occurs in other species. In the one species for which we know mental experiences do
occur abundantly, behaviorists have nothing to offer but the advice
that we should ignore them. Statements of interest or concern are
in a different category from statement of fact or statements about
the ease or difficulty of gathering a given sort of information. The
disinterest of one group of scientists should not inhibit others from
exploring aspects of the universe that they believe may be significant. There is no reason why behaviorists and sociobiologists
should not go their own way searching for external contingencies
affecting behavior, while those who find mental states of significant
interest do their best to study them wherever they may find evidence for their existence.

Without implying the existence of four-legged poets, six-
legged historians, or cephalopod philosophers, it may be appropriate to consider whether basic elements of what we mean by the
term *humanism* exist in other species. To the extent that ethologists
can learn what kinds of awareness and thinking may occur in various animals, we may hope to understand whatever value judgments accompany the flexible and versatile behavior of complex
animals. Altruism, for example, is receiving renewed attention
from sociobiologists. To be sure, their interest has so far been concentrated on its relationship to inclusive fitness, but equally im-

portant is the question of possible intentions and awareness of the likely results of altruistic behavior. Relevant evidence may be obtainable once ethologists take a renewed interest in the significance of the questions themselves.

Much bird behavior can be interpreted as resulting in enjoyment on the part of the performer or neighboring conspecifics. Examples include territorial singing when no rival males have been present for some time, the decoration of mating arenas (dramatically exemplified by the bower birds), and the many elaborations of courtship displays that appear to extend far beyond the immediate needs of sexual arousal and mutual synchronization of sexual behavior. Ethologists and sociobiologists have concentrated so heavily on the search for adaptive functions that they have tended to neglect (although seldom dogmatically to deny) the possibility that these activities produce something that can reasonably be called *pleasure* or *enjoyment* (Marshall, 1954). Hartshorne (1973) and Thorpe (1974) have expressed cautious recognition of the possibility that birds actually do enjoy their songs. Kroodsma (1977) has recently shown experimentally that tape recordings of highly varied songs stimulate female canaries more strongly than monotonous samples. It would be significant to inquire whether the females show any detectable signs of enjoyment as well as doing more nest building and laying more eggs. Such information, to the extent that it could be gathered in a convincing form, would complement, and would in no way contradict, the more conventional sociobiological interpretations concerned with evolutionary adaptiveness.

Available evidence provides only suggestive but inconclusive hints that some elements of humanistic qualities may exist in some animals, at least under certain conditions. But we can scarcely expect to obtain more conclusive evidence, pro or con, unless we begin to take such questions seriously and plan carefully designed scientific investigations in an attempt to answer them. There is a natural tendency to jump to conclusions, based on limited personal experiences with pets or other animals. But judgment should be reserved, and anecdotal evidence should be interpreted as suggestive, not definitive, for all the reasons that the early behaviorists emphasized half a century ago. Just as in other areas of biology, systematic, repeatable observations and experiments will be necessary before confident conclusions are warranted. But a necessary first step is to open our eyes to these possibilities and begin to adopt in ethology some of the approaches customary in humanistic scholarship—along with the continuing analyses of physiological mechanisms and evolutionary adaptiveness of animal behavior. Humanists are accustomed to dealing effectively with evidence that does not allow absolutely rigorous conclusions about causes and results. Cognitive ethologists can make good use of these kinds of critical scholarship, which have been successful in the humanistic disciplines.

The most serious reason for behavioral scientists to neglect questions about animal thinking is methodological. We can study responsiveness, discrimination between stimuli, and learning in animals, or we can analyze the appropriateness of behavior as an adaptation to various situations. But people also tell us about what they are thinking—at least what they believe or say they are thinking. This has generally been considered impossible in the case of nonhuman animals, for lack of language. This view is a special case of what has often been called *methodological behaviorism* (Lashley, 1923). Many ethologists and psychologists are quite open-minded about the possibility that animals may have mental experiences but see no practical way to study them. It will be helpful, before returning to this question in a later section, to consider which types of mental states might be most accessible to scientific scrutiny.

The terms *thinking* and *mental* are obviously very broad and diffuse—encompassing a huge variety of meanings. It is therefore appropriate as a first step to narrow the field of inquiry to a more manageable fraction of this vast and amorphous area. The general term *awareness* is generally understood to encompass such mental experiences as recognition, memory, anticipation, and intentions. If some kinds of awareness can be demonstrated in certain animals, a solid foundation can be laid for further investigations of cognitive ethology.

As pointed out by John Searle elsewhere in this volume, mental states or thoughts can conveniently be divided into two general categories: (1) subjective feelings and sensations that do not have obvious reference to external objects and events and (2) mental images that do pertain to something outside the organism in question. It may not be possible in all cases to maintain these distinctions in a rigorous fashion, but they are nevertheless very useful approximations. The philosopher's familiar toothache, the quality of redness, or feelings of hunger are in a real sense private, internal affairs, the subjective qualities of which are extraordinarily difficult to examine even in another person, let alone a member of a different species. The first type of mental state seems much more difficult to study in animals than are mental images that have definite reference to the world outside the animal.

I am using the term *image* in a broad sense to include any representation of the outside world of which persons or animals may be aware—that is, that they may think about and manipulate internally. It need not be a visual image but may be a pattern of remembered or imagined sounds, smells, or tactile perceptions. A mental image ordinarily resembles a sensation or perception except that it is not linked to current sensory input. Like perceptions, mental images are not static; they change with time, and one of their important properties is their temporal organization—that is, the pattern of their sequence, duration, and stability.

It is especially important to inquire whether and to what extent animals experience mental images of processes with a temporal dimension. If an animal communicates about its future intentions, such reports can be compared with what the animal actually does. Intention movements are a case in point; they communicate information about what the animal is likely to do in the near future. Of course, human words and sentences will be only rough approximations at best of any actual thought processes that might occur in animal brains, but imperfections of the postulated translation should not blind us to the likelihood that animals might experience thoughts roughly described in the following words: A hungry wolf: "If I chase that rabbit, I can catch it, and it will taste good." A ground squirrel: "If I dig this burrow deeper, I can crawl into a dark hiding place." A cottontail rabbit: "If I run into this briar patch, that big, threatening animal won't catch and hurt me." Or a male songbird: "If I sing loudly enough, that other male will leave my territory." Such plausible examples of "if . . . then" thinking can easily be multiplied by anyone familiar with animal behavior. The learned behavior of laboratory animals can equally well be interpreted in comparable terms such as (for a white rat), "If I press this lever, food pellets will fall out of that hole in the wall" or (for a pigeon), "If I peck when the light is red, there will be a loud clank, and I can reach some grain."

Animals often behave in ways that are consistent with the interpretation that they are thinking in such "if . . . then" terms. Predators stalk, pursue, capture, and eat their prey; rodents dig burrows and take shelter in them; laboratory animals learn new patterns of behavior that yield food, water, or other things they need and use. But, as behaviorists have emphasized for several decades, appropriate behavior does not prove that the animal has a clear understanding of what it is doing. Sleeping or anesthetized people can be conditioned, and machines, especially computer systems, have been devised that exhibit modifications of their activities that fit the formal definitions of learning. It is therefore customary to refrain from interpreting even the most versatile and adaptive behavior in animals as evidence that they are aware of anything at all.

I have recently pointed out that this prevailing view is based on negative evidence that in fact justifies only an open-minded agnostic position (Griffin, 1976). It is possible that not only people but also some animals are more than sleepwalkers.

Raising the question of possible mental states in other species sometimes encounters a special form of what I have elsewhere called the "so what objection." For example, Krebs (1977) and Humphrey (1977) vigorously assert that it would make no difference if we *did* learn that animals have mental experiences. This viewpoint is a most surprising one to find among ethologists whose scientific goal is to understand animal behavior. What could be more important information about a given animal than the fact that it was thinking certain kinds of thoughts, making particular plans, recalling such-and-such past experiences, holding one or another opinion, or wishing that something would or would not happen? The difficulties of gathering convincing evidence are formidable, and it is easy to understand that many ethologists may feel these difficulties are too great to justify the effort to learn about mental states in other species. But to insist that, even if such

information could be obtained in a satisfactory and wholly convincing fashion, it would make no difference whatsoever seems to
reveal a desire not to know certain things about animals even if
these things should turn out to be true.

Perhaps these views simply reflect an excessive concentra-
tion on other aspects of ethology—physiological mechanisms or
evolutionary background, for example. But such a narrow focus
is maladaptive, since it restricts the scope of potential insights. Neu-
rophysiologists who are not individually concerned about evolutionary origins of the animals they study are usually broad-minded
enough to recognize that such questions are significant even though
remote from their own work. Likewise, sociobiologists may have
little interest in physiological mechanisms, but they seldom say it
would make no difference whatsoever if a given behavior pattern
should be found to be based on a certain hormonal level or on a
particular sensory input channel or to result from endogenous
brain activity rather than from current sensory input of any kind.
The apparent desire to avoid any knowledge of mental states in
animals seems a most curious and paralytic frame of mind.

In order for us to detect and analyze awareness in another
organism, its images must be reported, under some circumstances
at least, through some type of observable behavior. Such reporting
must be capable of interpretation by another organism and of influencing its behavior in turn. Unless this reporting can occur in
the absence of contemporary stimulation that could directly establish a perception similar to a certain mental image, it is difficult or
impossible to distinguish between two possibilities: (1) The animal
may indeed be reporting about a mental image or (2) there may
be no mental image at all, and the animal may be reacting to a
perception of the current pattern of stimulation it is receiving. The
existence of intentions may be indicated by intention movements
or other signals, and the validity of their interpretation may be
tested by their actual predictive value. Does the animal actually do
what we predict from observing the intention movement? The far-
ther into the future such predictions can be extended and the
greater the difference between the situations when the intention
is expressed and when the intended action is performed, the

stronger is the evidence that an actual intention existed when the intention signal was emitted.

A widespread objection to the viewpoint just expressed is that the emission of information that permits prediction about future events is far from being sufficient evidence of conscious intention. For example, the squeaking of a worn bearing allows us to predict its imminent breakdown, and lightning flashes often permit accurate prediction of heavy rain. But we do not ascribe intent to the bearing or the cumulonimbus cloud. How, then, can we distinguish an animal's intentions from such simple examples as squeaky bearings and thunderclouds? One basis for such a distinction is, of course, the combination of firsthand knowledge of our own intentions, combined with recognition of the many similarities of brain and behavior between ourselves and other complex animals. We surely have more in common with mammals, birds, and even insects than with rotating machinery or with cloud formations. Another basis is the versatility and adaptability of communicative behavior in many animals, which certainly do behave in ways that are consistent with the interpretation that they have at least simple, short-term intentions. Both types of evidence are, of course, indicative rather than conclusive. An appropriate research objective for ethologists is to gather better evidence bearing on these questions.

Our mental images are often complex, versatile, dynamic, holistic, and multidimensional, rather than rigid, silhouettelike templates. Such properties make them more useful, since important objects can be recognized from new viewpoints or when only some of their properties are perceptible. Mental images can also be called into conscious recognition and manipulated in the absence of current sensory input. It seems likely that sensory input patterns are compared with such images to facilitate appropriate behavior when an adequate match is recognized.

Taking for granted the reality and significance of human awareness, we must recognize that we lack any neurophysiological methods adequate to detect physiological correlates of internal images. While hopefully awaiting a more adequate neurophysiology of awareness, we are therefore limited to inferring the presence of mental images and their attributes by observing behavior with which we suspect they may interact. Consistent actions are one type of helpful evidence that such images exist; for example, if persons or animals perform a learned response never observed before training, we infer some sort of memory trace within their brains. But we know from our own personal experience that mental images often exist in the absence of any overt behavior, and it seems beyond reasonable doubt that the same is true of our fellow men and women.

Human language provides abundant evidence about many kinds of mental images, although of course it is not totally complete or accurate under the best of circumstances, and it may be absent or totally false in the case of a person who is reticent, mistaken, or lying. Nonverbal communication also provides helpful evidence that fills in gaps or corrects errors that would occur if we relied entirely on the purely linguistic aspects of speech—for example, those that can be conveyed in writing. Perhaps, as I have suggested elsewhere, animal communication can also provide significant data about internal imagery in other species (Griffin, 1976).

A long-standing tradition in philosophy and linguistics stresses as a criterion for the presence of internal or mental images the ability of people to report about them even in the absence of contemporary sensory input from the corresponding objects or events. The operational usefulness of this criterion has led many philosophers and linguists to equate human thinking with the use of language. Hattiangadi (1973) has even argued that in our own evolutionary history the use of language must have preceded conscious awareness. If we make the simple extension of this operational definition to include nonverbal as well as verbal communication, we can also extend it to other species with little difficulty— unless we feel committed to a faith in human uniqueness for other reasons. If talking to oneself is an acceptable definition of thinking, as proposed by Skinner (1957), so is gesturing or "expressing" to oneself; and it is increasingly difficult to deny any measure of such communication to any other species. Indeed, if this approach is *not* reasonable, then we should reexamine the widely held view that human language and thinking are inextricably linked.

These considerations suggest defining awareness as the experiencing of mental images that can be reported. If this view is accepted, it is difficult to deny that some kinds of communicative behavior in animals serve this reporting function, since conspecifics as well as human observers clearly get the message and act accordingly (Sebeok, 1977; Smith, 1977). But we should recognize that important entities may well exist, even though we have no means of detecting them. Hence the operational definition just given would be foolishly restrictive if it required that the image-containing organism must report continuously about all of them. Some confidence is required that if it can report about them some of the time the images are also present at other times and in other circumstances. The alternative is to postulate that such images exist only during the act of reporting them, which is patent nonsense in the human case. While direct evidence is not available, it seems more likely than not that, insofar as reportable mental images also occur in animals, they also continue to exist between reports.

The possibility that many other complex animals besides the great apes and honeybees are capable of mental experiences certainly warrants further investigation. Monkeys, dogs, cats, and horses, for example, *may* know what they are doing; our problem as behavioral scientists is to find ways to gather definitive evidence. Whales and smaller cetaceans such as porpoises show abundant signs of complex mental abilities. They can learn to perform complex and even inventive kinds of behavior (Pryor, 1976); they communicate by a rich variety of underwater sounds; and their brains are as large as or larger than ours, with every sign of anatomical and physiological specialization for refined types of information processing. But no one has yet succeeded in decoding cetacean communication sounds in a way that demonstrates that messages of any subtlety are in fact being exchanged. This may be partly a matter of difficulties faced by an experimenter who seeks to observe the effects of a given cetacean sound on the behavior of a conspecific. From the data now available, we cannot conclude that the communication of whales is as symbolic—or as much about objects and relationships as remote from immediate stimuli—as the dances of honeybees. While, of course, we cannot be sure, it seems more likely that this situation results from difficulties in observing whale behavior and in correlating signals with responses, than that the rich repertoire of their songs serves only to convey such simple messages as individual or group identity. We badly need a von Frisch for the cetaceans.

Emergent Properties

A widely held view is that subjective awareness appeared in our own ancestors at some relatively recent point in our evolutionary history. Jaynes (1977) suggests, for the Greeks, a date between the times when the Iliad and the Odyssey of Homer were composed. Others would be more conservative and set such a date tens or hundreds of thousands of years earlier—perhaps even a few million years. There is no doubt, however, that some property that is essentially new can arise through gradual evolution, so that there is complete continuity over the course of evolutionary history.

A relatively uncontroversial example of such evolutionary emergence is powered flight, which has arisen at least three times among vertebrate animals and in several different groups of insects. While the exact details of the sequence are not known in any case, it seems likely, for example, that the ancestors of bats were small arboreal mammals that extended the range of their jumps by gradually improved forms of gliding. Intermediate stages in this process might resemble those attained by contemporary flying squirrels and flying phalangers, which can cover much greater distances by gliding than their size, weight, and muscle power would allow them to attain by simple jumping. They can even control their glide paths to some degree, pulling up at the end of a long glide to gain altitude while reducing forward speed. They can also steer horizontally and land in a reasonably coordinated stall that minimizes the force of impact. It is easy to imagine some gliding mammal extending the range of its glide by moving its limbs so as to create additional lift. Initially, this lift would be much less than the animal's weight, but a continuous series of small changes in structure and function could bring the animal to the point where its muscular exertions achieved truly sustained powered flight. Of course, even the strongest fliers eventually run out of stored fuel and water or encounter other physiological needs that force them to land. But small land birds migrating over the ocean can apparently remain on the wing for three days or more. (This elementary discussion leaves out the possibility of remaining aloft by gliding in rising air, which is important to many flying animals but clearly irrelevant to the general question under discussion.)

It is quite possible that conscious awareness as experienced by our species is an emergent property analogous to the sustained powered flight of bats, birds, and flying insects. Such a view is consistent with the "emergent materialism" advocated by Bunge (1977). If only one species of specialized flying animal existed—for example, the herring gull—and no intermediate forms were known, we might well consider its powers of flight qualitatively unique and totally different in kind from the locomotion of any other species. To the extent that such analogies have any validity, we might expect to find many kinds and degrees of awareness in various animals. We certainly know of no species other than our own that is capable of communicating and manipulating symbolic information with a remotely comparable scope and versatility. But this does not preclude the occurrence of similar, although probably far simpler, processes and experiences in other animals. Scientific exploration of this possibility and the attempt to obtain at least tentative answers to questions about the extent and nature of animal awareness provide significant and challenging questions for future investigations of cognitive ethology.

14 *Kenneth E. Boulding*

Sociobiology
or Biosociology?

The development of a new discipline is always an exciting event, and there is little doubt that sociobiology is a new discipline in biology. The Adam Smith of sociobiology, of course, is Edward O. Wilson of Harvard, whose monumental work, *Sociobiology: The New Synthesis* (1975a), is a vast mine of information. For those who cannot quite tackle this Matterhorn, David P. Barash's *Sociobiology and Behavior* (1977) will give an excellent overview of the field.

In its origin, sociobiology was an attempt—in considerable part successful—to expand the Darwinian (or perhaps one should say the neo-Darwinian) model of evolutionary dynamics to explain animal behavior as well as animal morphology. As animal behavior is, in effect, an aspect of this morphology in space and time, this certainly seems like a logical extension of the theory. Every individual animal can be thought of as a pattern in space and time, beginning with its genetic origins in the fertilized egg, divided cell, or injected virus and proceeding from there as a pattern of "growth," and also of behavior, which, like growth, is a pattern in space and time. The pattern ends, of course, with its eventual

260

death. Any general theory of evolution, therefore, would have to explain not only the form and structure of an animal at a particular point in time but also its whole life patterns, which would include its behavior.

The Darwinian model itself is an integrated structure involving a number of different parts. I am almost prepared to argue that it should be called a *vision* rather than a *model,* for as a model it is highly incomplete, mainly because of the enormous incompleteness and bias of the historical record. In a sense, the Darwinian model is an attempt to explain the total past of the universe, which is a very ambitious enterprise. This enterprise is seriously hampered by the fact that all our knowledge of the past is derived from those structures in it that have survived in the present, such as rocks; fossils; documents; archeological remains; patterns of radiation from distant sources, which are studied in astronomy; and regular breakdowns of such radioactive elements as carbon 14. Our knowledge of the past, therefore, is enormously biased by durability. The nondurable structures of the past simply do not come through to the present. We have a not wholly unreasonable prejudice that the durable is also the important, but we cannot be sure of this.

A good case in point is the evolution of spoken language, about which we know practically nothing, because in this century no records of spoken language survive except perhaps for a few generations in the memory of minstrels and taletellers. Since the invention of the phonograph, of course, we do have records of spoken language, but this is a very short period of time, in which spoken language has not evolved very much. Durability is a problem that biologists tend to brush under the rug. They simply assume enormous durability of genetic structures, for instance, and, while this assumption may be justified, the evidence for it is bound to be circumstantial and would not stand up too well in a court of law.

The two basic concepts of classical Darwinism are mutation and selection. These, however, are expanded in more modern theory to include two further concepts: *production,* which is the transition from the genotype to the phenotype, and *ecological interaction,* which is the major mechanism by which selection takes place. I am

prepared to define evolution, indeed, as ecological interaction under conditions of constantly changing parameters. The change in
parameters is mutation; the ecological interaction is selection. The
history of the universe can be written perhaps in terms of systems
of ecological interaction that are constantly moving toward an
equilibrium that itself is constantly moving because of mutation.
It is an endless succession of dogs chasing an endless succession of
rabbits that they never catch. Ecologists, like economists, tend to
have a touching faith in equilibrium. The awful truth is that equilibrium is a figment of the human imagination, although a useful
one, and that the real world has been a disequilibrium system ever
since the "Big Bang" or whatever it was that started it off.

    An ecosystem, or a system of ecological interaction, is a set
of populations of different species in which the rate of growth or
decline of the population of each species is a function of the existing size of all the populations, including its own. A *species* I define
as any set of objects that conform to a common definition and that
can be added to or subtracted from. This may include hydrogen
atoms; water molecules; quartz crystals; a particular strain of virus;
a particular kind of living organism; significant classifications of
human persons, such as blacksmiths; human artifacts, such as au-
tomobiles or spectacles; and human organizations, such as churches
or grocery stores.

    In a given environment of all other species, one can pos-
tulate a function relating the rate of growth in the population of
any one species to its own population. In its most general form,
this function is likely to exhibit a maximum rate of growth; as the
population of the species rises from low levels, the rate of growth
may increase to reach maximum, after which any further increase
in population results in a decline in the rate of growth—and at
some population the rate of growth will be zero, and at greater
populations it will be negative. The population at which the rate
of growth is zero is the "niche" of the species in that particular
ecosystem. If this point is at zero population or less, there is no
niche for the species in the system, and if the population already
exists in the system it will decline until it becomes extinct. If it is
not in the system already, it will never become a part of it.

    A concept that biologists seem to find difficult but that I

think is essential to the understanding of the dynamics of evolution is the concept of an "empty niche." This is the niche of a potential species that would have a positive niche if it were in the system but that has never come into it. Thus there was clearly an empty niche of considerable size for rabbits in Australia, and once they were introduced by human beings they expanded very rapidly into this niche, to the detriment of many of the existing species. Subsequently, a disease was introduced that severely reduced the niche of the rabbits. If there is an empty niche in an ecosystem, then there is some probability that it will be filled, either by genetic mutation or by migration. Empty niches, however, do not last forever. Mutation goes on constantly in other species or in the physical environment, and as this happens the niche may eventually close. If it is not filled before it closes, a certain opportunity is lost forever, and the whole course of evolution thereafter is different.

This is why evolutionary systems are not deterministic in the sense that an equilibrium system where evolution has stopped, such as the solar system, can be approximately deterministic. Evolution, in other words, is not a bit like celestial mechanics. It follows a very different model in which we are always coming to forks in the road and the dice determine which way we shall go. This does not rule out the possibility of convergent evolution—that is, where many different evolutionary paths may converge toward similar systems. A really good idea such as the eye, for instance, was produced by very different paths in the vertebrates and in the octopus. It may be indeed that there is, as Tennyson said, "one far-off divine event to which the whole creation moves" and that the whole evolutionary process in the universe is convergent toward some recreation of evolutionary potential. But this is poetic speculation; we have no way of knowing whether it is really true. Until we find out more about the long-run dynamics of evolutionary systems—perhaps not even then—we can never be sure whether the universe will end in a whimper or a thin thermodynamic soup or will evolve to the Omega point, as De Chardin (1961) calls it, of a new Big Bang.

We know more about the ongoing, day-by-day, year-by-year, age-by-age dynamics of the evolutionary process than we do about its ultimate end. In biological systems, the process is dominated by the distinction between the genotype and the phenotype—that is, between the egg and the chicken. Evolution is seen primarily as a process in the structure of what might be called the "geno-sphere"—that is, the sphere all around the earth of genetic infor-mation and instructions—witness Samuel Butler's famous remark that "A chicken is just an egg's idea for producing eggs." For bi-ologists, the mutation process takes place primarily in the geno-sphere, and selection takes place in the phenosphere—that is, in the totality of ecological systems of living species. The interaction of phenotypes is important in the selection of the structure of the genosphere. Genetic combinations that produce phenotypes that do not have a niche in some ecosystem will not survive. Those that produce phenotypes that have a niche in the ecosystem will survive. The phenotype also has the capacity to reproduce and to reform genetic material into new genomes, the genome being that partic-ular combination of genetic information and instructions that is the point of origin of a living individual.

In asexual reproduction, each individual, barring certain viral insertions of genetic material, has the identical genetic struc-ture of its "parent." Sexual reproduction apparently turned out to be a good idea—I do not know how many times it was invented; its origins are largely lost. But it did permit a species to rearrange constantly its pool of genetic information and instructions into new combinations and hence increased the variety of phenotypes of which any given gene pool was capable. This advantage evidently overcame the disadvantage of having to arrange for fertilization —that is, for the uniting of male and female genetic material. The extraordinary variety of devices by which this is accomplished in the biosphere suggests that the capacity of the variability in the phenotype introduced by sexual reproduction paid off extremely well in the finding of niches, although exactly why this should be so is not clear to me. Indeed, as one reads the evolutionary liter-ature, one constantly finds statements that imply, "this is the way it was, and you had better believe it," without any sense of the ne-cessity for its being this way. This perhaps reflects the fundamental indeterminacy of the whole evolutionary process.

Critical to the whole evolutionary dynamic are the processes of production by which the "know-how" that is contained in the genome (for instance, in the fertilized egg) is able to direct energy toward the transportation and transformation of selected materials into growing structures of the phenotype, beginning with the fetus. After emergence from the egg or the womb, the "know-how" of the whole organism utilizes the energy derived from the ingestion of food and water and the breathing of air to direct the materials (which also derive from food, water, and air) into the further growth of these improbable structures that eventually result in the mature animal or plant. As the growth of the phenotype proceeds, the structure of the phenotype itself begins to take on autonomous functions: The chicken breaks out of the egg, breathes, and looks for food; the newborn baby breathes, turns to its mother, and sucks milk.

Whether we have behavior before birth is an interesting, although perhaps not very important, question. What is certain is that from the moment of birth behavior assumes an increasingly important role in the survival of the individual. Part of this behavior is built into the structure of the phenotype by the genes. Once the genes have created the structure, even though they continue to cooperate with it and modify it, the structure itself exhibits autonomy; in other words, it behaves. Behavior is a realization of the potential of the structure that the genes have built and must be regarded as an integral part of the morphology of the individual living organism. At what point the behavior of the total structure of the phenotype takes over from the genes that are directing it is a very tricky question, the answer to which it is not easy to find. Certainly the genes that are still present in the body have very little to do with its behavior, although they have produced and continue to restore and to modify the structure that enables it to behave, whether this is in feeding, breathing, preying, fleeing, hiding, nest building, or mating. Indirectly, however, these behaviors affect the survival or nonsurvival of the genetic structure, because of the fundamental principle that a genetic structure that builds a phenotype that has no niche itself will not survive and will not be reproduced.

As the phenotype grows, it begins to use energy and materials for a number of different purposes. It will use energy, for instance, to sustain appropriate temperatures at which the physical and chemical and perhaps informational reactions that are necessary to its growth can take place most effectively. The chemistry of living processes seems to require the maintenance of a very narrow range of temperatures and seems to be most efficient in the neighborhood of our own body temperature, 37 degrees Celsius. As the genes build up nervous systems and means of communication, energy is also used to transmit information from one part of the body to another. This is extremely important in behavior, particularly in the more advanced forms of behavior that require transmission of information from some central nervous system or brain into the rest of the body. Information can also be transmitted chemically—for instance, through the blood stream—but this seems to be much less efficient than transmitting it electrically through nervous systems. A social parallel would be the contrast between the postal system and the telephone. The invention of the telephone had somewhat the same kind of effect on social organizations that the invention of nervous systems did on biological organizations, and it permitted a much larger scale of organization.

Once we get something like a genome with the capacity for producing a central nervous system or a brain, or even perhaps earlier than this, a new factor comes into the evolutionary process, which is *learning*. This is the transmission of some kind of information or know-how structure by direct information contact between older members of the species and the younger members. There is a change in the structure of the individual organism as a result of learning, which itself changes the behavior and affects the chances of survival and the size of the niche. Biologists often tend to underestimate this factor. Their "thing" is what might be called "biogenetics"—that is, the information and instructions obtained in the genome and in the genosphere—so that they are apt to overlook this factor, which also is transmitted from generation to generation, not through the genes but through the potential of the structures in nervous systems or even through other potential learning matrices that the genes have created.

There are essentially two genetic structures, if by genetic structure we mean a reproducible pattern of coded information capable of organizing the growth and behavior of products or phenotypes. The biogenetic process is coded in DNA and in the genes. It produces nervous systems (brains and the like) that have some biogenetically determined structure, from which we derive instinctual behavior, but the biogenetic process also, at some point in the evolutionary process, produces unstructured nervous systems that have a potential for receiving structure from information inputs and from self-generated information processes—that is, by learning. It is these learned structures that comprise the "noogenetic" process. (I confess to inventing the word, but it is badly needed.) Noogenetic structures and processes are just as "genetic" as biogenetic structures and processes. They are passed on from one generation to the next (usually with some change, but then there is also mutation in the genes). They are also capable of organizing behavior and even new species of artifacts. In the human race, the noogenetic processes are overwhelmingly dominant; biogenetics merely provides a vast unstructured brain, the eventual structure of which is produced almost wholly by noogenetic processes—that is, by learning.

Learning begins quite early in the evolutionary process. We do not really know how early. If the "worm runners" are correct, it begins even with the planaria, although there is some doubt about this. It is often very hard to distinguish the biogenetic from the noogenetic, simply because, in the lower animals especially, it is very hard to observe the learning process. I do not know whether an amoeba has a "culture"—that is, whether its genetic structure produces structures in the amoeba itself that are capable of change through contact with other amoebas or through its general environment. There is no doubt, however, that as we get up into the vertebrates learned behavior becomes increasingly important and that even many birds have to learn at least part of their songs. The genes produce a potentiality for the song that is not realized, however, if the bird never hears another bird of its own species sing. As we move into the mammals and toward the apes, the noogenetic factor becomes increasingly dominant, and in the human race, of course, it is completely dominant.

Behavior derived solely from structure in the phenotype that is built in by the genes is what is called *instinct*. Behavior derived from structure modified as a result of inputs of some sorts is learned. The distinction between instinct and learning is often very hard to draw and hard to test, simply because growing organisms tend not to survive if they are deprived of their normal environments. It is now becoming clear that the learning capacities of many species, particularly the apes and the monkeys, are much larger than we previously thought. The chimpanzees can be taught vocabularies of several hundred words in sign language, and Japanese monkeys can produce geniuses who change the behavior of all subsequent generations. It is an interesting question whether the skills of beavers at some point were not produced by some happy noogenetic mutation, no doubt of a beaver Frank Lloyd Wright who taught them how to build dams and houses, which knowledge has been transmitted from older beavers to younger beavers ever since. Or does the capacity of the beavers to build dams and houses depend entirely on structures built into their nervous systems by their own genes? I could see a large experimental science developing, trying to separate the biogenetic from the noogenetic by taking young away from their parents at the earliest possible age and raising them in various environments.

With the genetic changes that produced *Homo sapiens,* evolution on this planet went into a wholly new phase involving human learning and human artifacts, and the noogenetic aspects of evolution become dominant, to the point, indeed, where they may eventually displace the whole process of biogenetic evolution, which presumably produced the human race. That is, it is not inconceivable in the future that all living species will be human artifacts produced by deliberate genetic manipulation through noogenetic processes that are transmitted from one generation to the next continuously. There would be nothing "unnatural" about this, any more than there is anything unnatural about the development of life itself. It would simply be another phase of the ongoing evolutionary process of the universe.

Sociobiology has been accused of being racist, especially by Marxists. This accusation seems quite unjustified. We may perhaps accuse the sociobiologists of overemphasizing the biogenetic elements in evolution of societal systems and underestimating the noogenetic elements, but this has nothing to do with racism in the ordinary sense of the word. Racism assumes that the biogenetic structures of different groups of human beings differ far more than in fact they do. Biologists know very well, and so do the sociobiologists, that racial differences (things such as skin color and minor facial characteristics) are a very small part of the total genetic informational and instructural capacity of the human genome. My fertilized egg "knew how" to make a human male with a pinkish skin, blue gray eyes, and black hair. It did not know how to make one with black skin or with blond hair or with brown eyes. It did know how to make a heart, lungs, liver, limbs, skin, and brain— and I do not mind confessing that it knew how to make a pretty good brain; otherwise I would not be writing this chapter. My good brain, however, may have been the result of a quite fortuitous combination of genes from my father and my mother, and this fortuitous combination could take place with people of any race. The identification of an interest in the gene pool and in the genetic combinations of the human race with racism is simply absurd. Those who want to destroy human genetic research in the name of an imagined racism are no friends of the human race. Just as Social Darwinism was a profound misunderstanding of Darwin and the evolutionary process, particularly as it applies to selection, so racism in the accepted sense can only be based on a profound misunderstanding of the science of genetics. The only cure for misunderstanding, as far as I know, is further understanding.

Up to now, research in human genetics has identified some sources of genetic defects, such as sickle-cell anemia and hemophilia. It has told us very little about the genetic sources of genetic excellence and still less about any genetic sources of moral excellence or moral failure. It would be absurd to rule these out as inadmissable; if there is something to know about these things, we ought to know it. All human beings are not genetically identical; this means that they are not genetically equal, even though they are genetically very similar. We are learning furthermore that many genetic defects can be overcome through learning processes or through our knowledge of the physics and chemistry of the body. Some genetic defects in human capacity may result from genetically induced chemical defects that can be corrected. Even what seem to be genetic defects in learning capacity can be offset— and, one suspects, increasingly offset—by improvements in the learning process itself. We must recognize, however, that these improvements may not diminish inequality. They may merely raise the whole achievement level of the human race to the point where even the people who today are called *morons* can operate effectively and have reasonably satisfactory lives and the geniuses can use their genius even more effectively.

The political myth that all human beings are created equal is by no means a self-evident truth, in spite of the Declaration of Independence, and in a very real sense it is untrue. The proposition that all human beings come into the world with a very large potential, the greater part of which is likely to be unrealized because of defects in political and social institutions, would not only be a better political myth but it would also have the great advantage of being true. The fact that dominates the social system and that makes the noogenetic factor so much more important than the biogenetic factor in societal evolution is precisely the enormous redundancy of the human brain and the fact that—up to now, at any rate—we seem to have utilized so little of it. The human brain is like an enormous ballroom, into a small corner of which we all confine ourselves by learning not to learn. There are biogenetic variations in the ultimate size of the ballroom, but these are insignificant compared with the noogenetic variations in the processes by which we tend to use more or less of it. Changes in human learning techniques have had and can have an enormously large impact on social systems, far more than any conceivable change in the total genetic structure of the human population, barring some wholly unforeseeable increase in our capacity for genetic manipulation at the individual level.

Sociobiology, which is a perfectly respectable discipline within the field of biology itself, even if it does tend to underestimate the noogenetic elements in prehuman biology, should not be confused with what might be called "biosociology," which is the illegitimate use of biological analogy in social systems. Unfortunately, there is a good deal of this, particularly in works of what might be called "pseudoscientific journalism," such as those of Robert Ardrey (1966). Even biologists of distinguished accomplishment in their own field, such as Konrad Lorenz (1966) or Lionel Tiger (1972), are apt to make serious mistakes when they try to generalize from biological systems into societal systems that are much more complex. There are some important similarities between biological and societal evolution, but these similarities should not blind us to the enormous differences, and each system must be analyzed on its own merits and according to its own peculiar properties and principles. It is particularly illegitimate to assume, as Robert Ardrey does, that because certain distant relations of the human species in the animal kingdom exhibit group territoriality and male aggressiveness these modes of behavior must be imprinted by our genetic structure in the brains of the human species.

Territoriality is a very legitimate subject for sociobiology and a very interesting one. It represents one of the many strategies of evolutionary survival, of which nonterritoriality is another. There are probably just as many, if not more, nonterritorial species as there are territorial ones, simply because any ecological system provides an enormous variety of niches and each niche that is occupied creates others of a different kind.

We do not even know how much behavior, especially in the higher animals, is noogenetic and how much is biogenetic. One suspects that a lot more is noogenetic than has previously been thought and that, in the absence of any very good techniques to separate the two, there is a strong tendency among biologists to assume that almost everything is biogenetic. Even within the close animal relatives of the human race there is a wide variety both of territoriality and of behavioral pattern from the highly territorial, ganglike baboons to the easy-going and gentle gibbons. Both patterns could probably exist among the direct ancestors of the human race, as both patterns are found in human culture.

The enormous cultural differences among peoples who are genetically very closely related, as observed by Margaret Mead and many other anthropologists, is striking testimony to the fact that the biogenetic element in human culture is quite small and that noogenetic elements are overwhelmingly important. Noogenetic evolution, however, like biogenetic evolution, has strong elements of indeterminacy. The accident of a natural disaster, or military defeat, or the unusual character of the leading person in the culture can direct it along certain lines and can cause it to diverge almost indefinitely from similar cultures that had other accidents happen to them. A striking example of this are the extraordinary differences among the courses taken by three East African countries—Kenya, Uganda, and Tanzania—which were part of the

British Empire until the early 1950s but which have followed completely divergent paths in the last twenty-five years and are likely to continue to do so for a very long time, so that they will end up being very different from each other, whereas twenty-five years ago they were similar in many respects. This is a result of the accident of the personalities of their national leaders, which produced an uneasy and rather corrupt capitalist development in Kenya, a much more stagnant socialist development in Tanzania, and a tyrannical retrogression in Uganda.

It would not be an unreasonable conclusion to suppose that the impact of biological sciences on the social sciences has been for the most part very adverse, occasionally almost catastrophic. We see this, for instance, in the late nineteenth-century Social Darwinism of Herbert Spencer in England and W. G. Sumner in the United States, which was guilty of two basic fallacies. One derived from Darwin's extremely unfortunate metaphor of "struggle for existence," which conceived evolutionary survival in terms of aggressiveness and a kind of machismo, whereas in fact struggle of any kind is very rare in biological evolution, where ecological interaction and survival occur through innumerable types of behavior and where in fact the lion is a much more endangered species than the mouse. The best examples of struggle in biological evolution are in sexual selection, where the males fight, or at least compete, for the females. This frequently leads to a line of development that leads ultimately to extinction, more rapidly than does the development of meeker and less conspicuous species. The meek—certainly the adaptable—do seem to inherit the earth, as the aggressive and the combative seem to kill each other off. Even in societal evolution, nondialectical and even nonconflictual processes are by far the larger part of the total dynamic, even though there are times when the outcome of a struggle can determine in part the future evolutionary pattern of the system.

The second fallacy of Social Darwinism was its identification of societal evolution with social laissez faire. The parallel, indeed, is of interest. An individual organism is a planned economy, with its plan in the genome (fertilized egg) and the living organism itself representing a fulfillment of the plan. Every fertilized egg is a plan for making a particular kind of organism, not any other kind.

However, an ecosystem is a riotous, anarchistic, laissez faire economy. It may have a dominant species that sets the niche pattern of many others, as for instance the pine does in a pine forest, but there is no government, no ruler; the dynamics of the system simply rest on the fact that the rate of growth of each species, as we have seen, is a function of the populations of all of them. This gives a potential for a kind of Walrasian equilibrium (Walras, 1954), in which all populations are stationary, the food chains cycle throughput of material without change of population of any species, and there is even a kind of equilibrium price system in terms of "the distribution of trade"—what each species gives up and takes out from the system. There is even something like the GNP in terms of the total biomass and the total throughput through it. If there is a large inflow of energy from the sun, both the biomass and the throughput are likely to be larger, just as the GNP is strongly relatd to the flow of energy through the system.

These parallels, while they are instructive, do not imply that biological ecosystems and social systems are identical. Social systems contain human beings who have consciousness, who have images of the future, who have very complex systems of values; who have "know-what" and images of abstract systems, as well as "know-how"; and who also have political institutions and governments that in some sense represent a "general will." Social systems, therefore, must be treated for what they are, and particularly it must not be assumed that they are simpler than they are. They demand their own description and their own evaluations.

Thus societal evolution is much more teleological than biological evolution. People have complex images of the future, some of which are more highly valued than others, and the selective process is strongly biased toward the higher values. Even though we do not always reach the future that we value most highly, the fact that we do put different values on alternative potential futures changes the future itself of the whole system. In biological systems, we have the beginnings of teleology in animal behavior, as for instance when a predator pursues its prey or when even the simplest animal accepts food and rejects nonfood. But these occur only in microsystems. No living creature that we know of before the human ever had an image of the system as a whole. The human spe-

·cies has been able to do this at least in part because of its capacity for language and for images of things that are beyond the personal experience. I doubt if even the most intelligent and linguistically well-trained chimpanzee could be taught geography.

Nevertheless, in spite of these differences, it is entirely legitimate to regard societal evolution as a continuation of the larger evolutionary process in a new phase. One of the great principles of evolution is that evolution itself evolves. The patterns of pre-biological evolution that produced the elements, the compounds, the rocks, the mountains, the oceans, and the atmosphere are much simpler than the processes of biological evolution that began with the development of DNA, a self-reproducing set of realizable instructions. Even within biological evolution we see changes in the pattern with the invention of sex, the vertebrate skeleton, the developing brain, and so on.

Societal evolution is largely the record of the development of human artifacts, which are products of human know-how, just as living organisms are the products of biogenetic know-how. These include material artifacts, of which there are now more species than there are living organisms. Human artifacts also include human organizations, from the family, the tribe, through to the church and the corporation, the state, the university, and the commune. Human persons likewise are in a very large measure human artifacts. Each of us is in part a biological artifact of the know-how in the fertilized egg, but the language that we speak or write or read, a very large proportion of our total behavior, our professions, our skills, and the roles that we occupy are to a very large extent made by other human beings or by the artifacts of other human beings, in books, instruction manuals, grammars, blueprints, and so on.

All these human artifacts interact with each other and with the species of the biosphere and the species of the physical environment in a very complex ecosystem that selects those species for which there is a niche, while others decline to extinction or are stillborn even if a mutation produces them. New species of artifacts are continually being created by mutation. This originates in human knowledge and know-how, which is the genetic structure of societal evolution. They are produced much as biological species are produced by the application of this know-how in the capture of energy for the transportation, transformation, and rearrangement of materials; the sustaining of temperatures; and the transmission of messages. The basic field of societal evolution, as of biological evolution, is the genetic structure, the know-hows that know how to produce an artifact. This know-how is coded in energy and materials, the absence of which may limit the coding and may limit the evolutionary process through the inability to produce artifacts—that is, phenotypes—that will transmit and expand by a selective process the overall genetic structure. Thus the absence of energy at the poles certainly limited both biological evolution and societal evolution. Energy and materials, however, are limiting factors; they are not the field within which the evolutionary process takes place.

The evolutionary approach to social systems has important implications for social policy and for the evaluation of social movements. In the first place, it repudiates the concept of equilibrium and sees the world as a continuing, ongoing flux that never repeats itself, although it exhibits patterns, on which endless variations are played. It is a little unfriendly to those apocalyptic and revolutionary views of society that seek a single, once-and-for-all solution of social problems. It is agnostic about ultimate objectives and final solutions, but it does not hesitate to evaluate social systems and to try to evaluate "betterment"—that is, whether things are going from bad to better or from bad to worse. It recognizes the enormous complexity of the ongoing dynamic of the total system of planet Earth, and it recognizes that because of this complexity, especially because of the large number of ecological relationships involved, "counterintuitive" results of human action are frequently observed. Many of the things we do to hurt people actually help them; many of the things we do to help people actually hurt them. Decisions are made among imaginary futures in unrealistic worlds, and the results of decisions are very frequently very different from what the decision maker imagined when the decision was made. In the light of this complexity, there is a constant temptation to fall into despair or into apathy. It was Malthus, the first evolutionary social scientist and a great moral theologian, who remarked, "Evil exists in the world not to create despair but activity" ([1798] 1959,

p. 138). And beyond the temptation to despair there is a very solid optimism. The world will change regardless of what we do about it. And we can put in our two bits' worth to divert the course of evolutionary change, as our understanding of it grows, toward the better rather than toward the worse.

15

George Wald

The Human Condition

$\mathcal{W}$hat impresses me most about the new sociobiology is its massive irrelevance to the present human condition. It is a little late to begin tracing our present state and its problems back to genetic sources. The genes are only permissive. We exploit our genetic potentialities less and less as we assume more and more specialized roles in more and more urbanized, mechanized, and industrialized societies. The limiting factors in modern human performance are primarily social, economic, and political.

We do indeed have an animal heritage of behavior, but by now large portions of it have become inappropriate to civilized life. This is true of the violent emotions—pain, fear, rage—as Walter Cannon remarked many years ago (Cannon, 1927). These emotions accompany bodily changes—the adrenal syndrome—that prepare one to fight or run. When neither action is carried through, they leave one sick, tense, jumpy, and with heart pounding. The violent emotions are out of place. A civilized person cannot afford them and should restrain them.

Similarly, the great human experiences of birth and death are rapidly being removed from our experience. Death is treated impersonally; it is hidden, external, a problem in sanitation. And few mothers in the upper strata of our society experience the births

277

of their own children, let alone the births of other women's children.

We are taught from early childhood a strangely distorted view of what is called *human nature*. I have never heard that term used except in disparagement. It is like what we call *realism* in literature: all misery, brutality, and blood on the floor. I have always been told that my neighbors want only to tear me apart and go off with all I have and are restrained from doing so only by the police. However, that is not my experience. On the contrary, I find my neighbors gentle, generous, cooperative, and not at all aggressive, much less violent. What I have to worry about is the police.

For some ten thousand years, human beings lived in relatively small groups, in a village culture that was tolerant of life: human, animal, and plant. Hardly two hundred years ago, what became the "developed" world launched out on a new kind of enterprise, the Industrial Revolution, which is now devastating our planet and threatening human and much other life on earth with extinction. It has put the human enterprise on an exponential growth curve. At the bottom of the chart, one writes the years, although the way things are going it hardly matters what happened much more than a century ago. Up along the side of the chart, one writes all kinds of things: population, industrial pollution, the use of fossil fuels, the consumption of other irreplaceable resources, armaments—and information.

We are living in an information explosion that is forcing modern scientists to keep narrowing their fields in the constantly frustrated hope of being able to keep up. That exponential curve is reaching for the moon in all those aspects at about the same time, close to the year 2000. I am one of those scientists who, try as I will to evade so calamitous a thought, cannot see how the human species is to get itself much past the year 2000—unless we change our present directions, unless we get that exponential curve to flatten out and in some respects to bend downward.

None of that has much to do with genes or with our animal heritage. It is all cultural, all socioeconomic.

The recent surge of industrialization and urbanization has created great hordes of helpless people, wholly dependent on finding "jobs," with no means of support except working for others. And with the increasing mechanization, even of agriculture, all over the world, there are constantly fewer jobs—more people and fewer jobs.

But there is a new phenomenon much worse than joblessness. Robert McNamara, ex-Ford Motor Company executive, ex-Secretary of Defense, now President of the World Bank, in his report to the World Bank in September 1970, spoke of "marginal men." "Marginal men" are persons who are not just out of work but also for whom in a market economy there is no further use, either as workers or customers. They are just a burden on the economy, an embarrassment, a potential source of trouble. McNamara estimated that in 1970 there were about 500 million such persons in the world—about twice the present population of the United States—and that in 1980 there might be one billion, and in 1990 two billion; that would be about half the present world population.

This situation is all due to biology, since it is a product of human activity and it is all ultimately part of natural selection in some broader sense. But it is biology that is doing violence to life, that is distorting and even reversing much that happened before in evolution.

The greatest event in the history of life on this planet was the development of photosynthesis. Before that, life had to feed on organic molecules—carbon compounds—accumulated over past ages from a variety of geological processes, mainly in the upper atmosphere and continuously leached out of the atmosphere into the oceans, in which life eventually arose some 3 billion years ago. With only that process to feed on, life would necessarily have had to end when the supply of organic molecules ran out. Fortunately, before that could happen photosynthesis developed—the process by which plants make their own organic matter from carbon dioxide, using the energy of sunlight. Through photosynthesis, life on earth was freed from its dependence on past accumulations of fossil organic matter and could take off on its own.

Our new, highly mechanized agriculture—the methods of agribusiness, the Green Revolution—has reversed this development. The photosynthesis that produces our food has now again been made dependent on fossil organic matter, on fossil fuels. A recent estimate is that the United States alone uses the energy equivalent of more than thirty billion gallons of gasoline per year for agriculture (Cloud, 1973, pp. 9–11). That makes a poor bargain, for that input of fossil fuels is estimated to represent about five times the energy content of the food that reaches our tables.

To go with this bad biology, there is a strange new sociological development—a sudden, explosive demand for meat throughout the developed world. Meat, particularly beef, has become the measure of the standard of living. Americans—other than our poor, both rural and urban—have enjoyed diets high in meat throughout our history; but recently the USSR, Japan, and Europe have reached the same level of meat consumption that the United States attained in 1940.

This new demand for meat means, among other things, that the grain that might otherwise feed hungry people in our country and in other parts of the world is used instead to feed pigs, sheep, and cattle. This is another poor bargain, for it takes up to eight pounds of grain to yield one pound of meat. Borgstrom (1974) of Michigan State University has estimated that the U.S. livestock population consumes enough food materials to feed 1.3 billion people. The world livestock population consumes foodstuffs that could otherwise feed about 19 billion people (Cloud, 1973).

Human biology has come to a strange pass, but this pass has little to do with our genes or with our animal heritage. It is uniquely ours, all new and all of our own making.

I have heard the silly question asked, "Why is our world five billion years old?"—and the silly answer, "Because it took that long to find that out!" The very science that has found that out has put into human hands the means of our own self-extinction. Some years ago, the stockpiles of nuclear weapons in the United States and USSR reached the explosive equivalent of about fifteen tons of TNT for every man, woman, and child on earth. For the past six years, the United States has been making three hydrogen warheads per day—about one thousand a year; and the USSR, of course, keeps pace with us. Meanwhile, other nations have gotten into the act. The so-called nuclear club now numbers six nations. With the spread of nuclear power and its by-product, plutonium 239 (the most effective material for making fission bombs), the nuclear club threatens to expand to as many as twenty-five nations within the next decade.

Our society has turned lethal, even self-destructive. Arms now represent the biggest business on earth, now at about $340 billion per year. I am trying to learn what the nuclear weapons business—that is, warheads, guidance systems, and vehicles—is worth in this country. My friends have suggested perhaps $10 billion per year. If this estimate is correct, the nuclear weapons business is a big, lucrative, and well-concealed business.

It is curious that E. O. Wilson (1975a, p. 552) speaks of money as "a quantification of reciprocal altruism." It happens here to quantify a special form of reciprocal service, with the official name *mutual assured destruction* and the interesting acronym MAD.

The new sociobiology appears just now as a strange distraction pursued in a society "on the brink," having already stockpiled all the nuclear hardware that could finish it off in a half hour's major exchange between the superpowers. Sociobiology is, perhaps, not an altogether irrelevant distraction, but it is potentially part of the story, namely the end. Some of my friends are starry-eyed about the prospect of establishing radio communication with more advanced technological societies in outer space and perhaps entering in that way an already existent galactic network. They have been listening for meaningful signals now for a generation, without result. The question has begun to be asked, "*Are* there more advanced technological societies in outer space? Or do they self-destruct, as we are threatening to do, once they reach approximately our stage of development?"

Let me say now that I see all these frightful things in our present reality, but I accept none of its portent of doom as inevitable. It is just that, with astonishing suddenness, it has become very late. That is the way exponential curves work—and there is no time to lose.

What can we do? Middle-class intellectuals have a ready, indeed a reflex, answer: "Do research." Research is fine; we never know enough about anything. But research must not be allowed to be used as a trap, as a means of endlessly putting off action.

We already know enough to *begin* to cope with all these

problems that now threaten our lives and the continuance of the human enterprise. Our crisis is not one of information, but of decision, of policy. It is one of our prevalent myths that government policy is based on the best information available on what would be most advantageous to the welfare of our people and our environment. The reality is usually just the opposite: Policy does not follow information—information follows policy. The government decides policy on economic and political grounds and then seeks out, purchases, or cadges, in one way or another, the information needed to support it.

We already know enough to begin to cope with all our most threatening problems. But in our part of this politically divided world we cannot begin to cope with any of our problems if we insist on maximizing profits. In some parts of the so-called socialist world, it is not very different—for I think we are beginning to realize that societies organized to maximize production, whatever their politics, end up not very different from societies such as our own, which is devoted to maximizing profits.

Surely the scientists' role in such a time as this is not merely to observe, measure, and record—to study nature as it goes down the drain, to study human societies as they are degraded and threatened with extinction. Society delegates to scientists, as it once did to priests, its grip on reality and its need to be preserved from the forces that threaten it. That is part of our social bargain, of what it must mean to be a scientist. A scientist should not just study nature but should take care of humanity, life, and our planet. We do not have the power to do those things by ourselves, but we are in the best position to know and learn what most needs to be done and to move our fellow human beings in those directions.

Michael S. Gregory

Epilogue

$\mathcal{A}$s Heraclitus said, "Nature loves to hide." It is probably equally true that human beings tend to hide from nature, at least whenever a scientist proposes links between the animal and human worlds that are too close for comfort. The first scientist to argue systematically that human beings are in fact animals and are subject to the same evolutionary processes was Charles Darwin, in his *Descent of Man* ([1871] 1969). As Edward O. Wilson observes in the introduction to this volume, sociobiology is an offspring of neo-Darwinian theory. Just as neo-Darwinism itself was the result of combining the evolutionary hypothesis with Gregor Mendel's discovery of the particulate basis of inheritance, sociobiology is the result of combining population genetics (which represents the culmination of neo-Darwinism) with ethology (the study of behavior). Each step of this development has made it increasingly difficult for human beings to hide from nature, especially from their own inherited nature as animals, as highly elaborated members of the primates, and as near or distant kin of every animal that has lived on this planet.

Sigmund Freud once remarked that history is a series of blows to human narcissism: For example, consider Copernicus' discovery that Earth is not the center of the solar system, Darwin's

283

discovery that we are descended from earlier animals, and Freud's own discovery that human behavior is often motivated by unconscious forces. Sociobiology would surely qualify as a fourth blow to our narcissism, for it contends that not only human structure but also human behavior and even consciousness are shaped by genes.

The consequences of such a theory are, of course, enormous. Probably the most profound of these is the recognition that in addition to our being descended from earlier animal forms the behavior of these earlier animal forms also survives within us. As this book demonstrates, great controversy surrounds the extent of this survival. What human behaviors can truly be said to be genetically determined, without resort to metaphorical terminology or questionable analogies? To what extent and under what circumstances can these behaviors be expressed, modified, limited, or even canceled by the operation of culture? Is culture itself genetically acquired or predisposed? If so, is not the genes-culture distinction a false dichotomy when considered from the standpoint of evolutionary biology? If the dichotomy is not altogether a false one, how are human beings to manage the often conflicting demands of gene-derived impulses toward behavior on the one hand, and the normative or prescriptive modes of behavior derived from culture, on the other? As Wald (Chapter Fifteen) puts it, "We do indeed have an animal heritage of behavior, but by now large portions of it have become inappropriate to civilized life."

Let it be noted once again that if the sociobiologists are proven right we human beings are in the grip of forces not only older than our civilization but also older than *Homo sapiens* and thus well beyond our control. If that is so, the Darwinian and Freudian blows to our narcissism in fact become a single, staggering blow, and we must begin all over again to understand ourselves in terms of the source and nature of our actions.

What are genes, that they should have the power to determine our bodily form, our behavior, and even our thought? Without going into the structure and dynamics of DNA and RNA, let us consider this passage from Richard Dawkins' *The Selfish Gene* (1976, p. 37): "Individuals are not stable things; they are fleeting.

Chromosomes, too, are shuffled into oblivion, like hands of cards soon after they are dealt. But the cards themselves survive the shuffling. The cards are the genes. The genes are not destroyed by crossing over; they merely change partners and march on. Of course they march on. That is their business. They are the replicators, and we are their survival machines. When we have served our purpose, we are cast aside. But genes are denizens of geological time: Genes are forever."

Dawkins goes on to say that individual genes, as biochemical entities, are not themselves immortal, but their replicas are. Of course, some genes do die if they seriously limit the reproductive fitness of their hosts; and some genes change as the result of mutation induced by radiation and possibly other means. But in the main, genes are as immortal as anything about which we have knowledge. The gene, as Dawkins (1976, p. 36) puts it, "leaps from body to body down the generations, manipulating body after body in its own way and for its own ends, abandoning a succession of mortal bodies before they sink in senility and death."

In fairness to sociobiology, Dawkins' view of genetic determinism is the most boldly reductionistic one published thus far. Many sociobiologists presumably would shudder at the suggestion that we are mere "survival machines" for the genes we inherit and pass on to our descendants. Although there seems little doubt that genes influence our behavior, it is unlikely that they do so in any direct and immediate way. For human beings, at least, the intervening variable between gene and behavior is culture. Culture is the principal accomplishment of *Homo sapiens,* and it remains the one fairly durable distinction between us and the other animals.

Thus the sociobiology controversy centers on the role of culture in human affairs. The definition of "culture as learned behavior" used to be universally accepted by anthropologists. Obviously, if behavior were genetically preconditioned it would not have to be learned. That definition has become blurred in recent years, as Washburn (Chapter Three) and Boulding (Chapter Fourteen) note, because we have discovered that learning plays a far greater role in animal behavior than we had expected. At the same time, the plasticity of human behavior, considered across cultures as well as through time, and above all the learned acquisition of language in all its varieties, mark the major divisions between human and animal behavior.

One of the problems sociobiology encounters in seeking genetic determinants of behavior is that it must explain everything or else it explains nothing. If there exists in a single species a single behavior that is intrinsically incapable of explanation on genetic grounds, sociobiology drops from a universal to a limited hypothesis. If the species in question is *Homo sapiens,* then sociobiology, in Holton's terms (Chapter Four) drops from a General Discipline to a Special Discipline. Students of sociobiology have therefore been forced to postulate that culture, in all its otherwise unmanageable variety, is itself genetically predisposed—that there are, in effect, *genes for culture.* Sociobiologists prefer to substitute the word *social* (see Wilson's introduction to this book), because social behavior is well known in nonhuman animals from insects upward, and there can be no argument that social behavior among the Hymenoptera, for example, is not hereditary. The terminological reductionism involved in replacing *cultural* by *social* effectively begs the question of the differences between humans and animals.

Even the staunchest sociobiologists do not argue for a one-to-one correlation between gene and behavior. They recognize, along with Barash (Chapter One), that behavior is a transaction between the gene and the environment. In the literature of sociobiology, behavior is often termed a *phenotype,* and the maxim "Genotype plus environment equals phenotype" is generally accepted. One of the unrecognized, and hence unresolved, controversies within sociobiology centers on the term *environment.* In some instances (for example, see Barash, Chapter One), environment appears to be taken in its conventional and literal sense, that is, as *physical surroundings.* Thus the genotype transacts with the environment—weather conditions, terrain, food supply, competitors, and predators—to produce the behavioral phenotype. In other instances, however, environment is used to describe the *social setting* (Wilson, Introduction) with which the genotype interacts to produce the behavioral phenotype. But the social setting itself is the product of sociobiological forces (in Wilson's view) and thus is far from being equivalent to physical surroundings. It is my intention not to resolve such terminological controversies but merely to point out that they probably indicate deeper methodological problems within sociobiological theory.

It is also instructive to watch strategic shifts of terminology within the course of a single argument. In outlining what he sees as the three possibilities for genetic determination of behavior, Wilson (Introduction) uses the word *culture* only in connection with a position he apparently regards as untenable: "that the human brain has evolved to the point that it has become an equipotential learning machine entirely determined by culture." In describing the other two positions, the first of which he regards as marginally acceptable and the second entirely so, Wilson substitutes the word *social*. Social evolution, as we have noted, is friendly to the sociobiological hypothesis, while cultural evolution is not. The reason for this is supplied by Barash (Chapter One) when he remarks that "unlike organic evolution, cultural evolution is Lamarckian, not Darwinian," and hence beyond the reach of sociobiological explanation.

It would seem futile for sociobiology to ignore the reality of culture and to go on trying to resolve the gene-culture dilemma by substituting *social behavior* for *culture*. Washburn occupies the high ground of empirical fact when he observes (Chapter Three) that "starting a little before agriculture, there was a very rapid change in technology that spread all over the world in a very few thousands of years. The conditions that had dominated human evolution for some millions of years changed; the nature of the change, the speed of the change, and the rate of diffusion all show that the changes were the result of learning, not of biological evolution." Barash (Chapter One) understands the limits of sociobiological explanation without attempting to dismiss the problem by shifting terminology. He notes that culture itself has evolved, but by processes only analogous to those of biological evolution, and recognizes that "the results of these two very different phenomena can be disquietingly similar and will be a challenge [for sociobiology] to assess."

A second major area of controversy in discussions of sociobiology is the question of mind. This question is still very much alive almost 120 years after the publication of *On the Origin of Spe-*

cies ([1859] 1964), in which Darwin excluded the subject from his general theory of evolution. Darwin's dispute with Wallace and the Wells-Shaw lament that Darwin had "banished mind from the universe" are too well known to be repeated in detail here. Only within the past decade has there been a marked revival of interest in mind, consciousness, and intelligence and in their relation to evolution. Some of the reasons for this long eclipse may be found in Holton's discussion (Chapter Four) of the vitalism-mechanism controversy and the triumph of mechanistic determinism in twentieth-century science. Mind and consciousness are insubstantial and not susceptible to measurement; intelligence can be measured, but only in rough approximations. Everything about the modern scientific paradigm has militated against serious investigations of such metaphenomena as awareness, thought, feeling, and intuition.

The authors represented in this volume are as sharply divided on the question of mind as they are on the question of culture. Pierre L. van den Berghe (Chapter Two), for example, sees human self-consciousness as a fetish used by social scientists to ward off intrusions by biologists seeking to explain human social behavior on genetic grounds. Yet Marjorie Grene (Chapter Eleven) accuses evolutionary biology of substituting explanations of the mind's origin for explanations of its function: "Sociobiology . . . purports to answer philosophical questions about mind by specifying necessary biological conditions for its evolution . . . to discover the biological conditions for mental development is not to say how, within those conditions, mind works." Boulding (Chapter Fourteen) considers culture and mind as parts of a single noogenetic process that has taken "evolution on this planet . . . into a wholly new phase" and warns sociobiology not "to underestimate the noogenetic elements in prehuman biology."

Wilson (Introduction) reviews the problem of mind and identifies "mentalism" as the foremost antagonist to sociobiology. "The human mind, this argument often goes, is an emergent property of the brain that is no longer tied to genetic controls. All that the genes can prescribe is the construction of the liberated brain [whereas in fact] models have already been produced in neurobiology and cognitive psychology that allow at least the possibility of mind as an epiphenomenon of complex but essentially conventional neuronal circuitry." If Wilson detects a difference between his "mind as epiphenomenon" and his opponents' "mind as emergent property," he does not choose to share this difference with us. I suspect that the hidden controversy is that between vitalism and mechanism, the history of which has been admirably outlined by Holton (Chapter Four).

Our other authors are wary of the jaws of this ancient dichotomy and steer well clear of it. Fuller (Chapter Five) recognizes the difficulty of explaining individual brain development as "the action of a batch of DNA and RNA molecules on a collection of amino acids," when in all probability "the genome does not contain a blueprint for the brain." His conclusion is that the brain is genetically programmed not for a specific development but for a readiness to develop in manifold ways. "The unique quality of the human genotype," he says, "is that it has guided the development of a brain with many options, each as natural as any other." This is tantamount to saying that the structure of the brain is genotypic, while its neurological development is phenotypic. If this is in fact Fuller's claim, it would seem to stretch sociobiological genetic determinism to its breaking point. It is precisely this argument that Wilson dismisses as "mentalism."

Searle (Chapter Eight) fails to detect anathema in mentalistic accounts of behavior. On the contrary, he finds that "the explanation of large areas of human behavior and presumably large areas of other animal behavior is ineliminably intentionalistic in form" and is therefore "not deterministic." Searle's hypothesis about animal behavior is based on the assumptions that animals are aware of present circumstances, that they are aware of (remember) similar circumstances in the past, and that they can form mental images of circumstances that do not presently exist but that can be made to exist in the future if the animal performs certain actions. These mental images Searle calls *intentionalistic representations,* and he finds that the existence of such will-driven mental imagery is the only logical inference that can be drawn from the observation of certain classes of animal behaviors. Although Searle himself cautiously avoids the term *consciousness,* his argument favors the existence of at least some level of consciousness among animals: "The higher animals . . . are capable of forming representations of their future actions and of then acting on those representations. One of the many questions this poses for the sociobiologist is, 'What sort of neurophysiological mechanisms—and what sort of evolutionary development of those neurophysiological mechanisms—made intentionality possible?'"

These are exceptionally important questions, but sociobiology is unprepared, at least at present, to answer them. Sociobiology is founded on thoroughly deterministic assumptions about behavior. Searle argues that some animal behaviors are nondeterministic. Sociobiology posits that behavior is a result of genetic and environmental forces and therefore is essentially passive response. Searle argues that behavior can be active, that it can create new circumstances rather than merely respond to existing ones. Again, Searle's discussion leads us in the direction of consciousness; if an animal is aware of the present, can remember the past, and can imagine a future—that is, form a mental image of something that does not yet physically exist—it is indeed difficult to maintain that the animal is not in some sense *conscious.*

Griffin's discussion (Chapter Thirteen) parallels in many respects Searle's. For example, Griffin speaks of "endogenous brain activity" and of mental imagery in humans and animals that is not determined by external stimuli: "I am using the term *image* in a broad sense to include any representation of the outside world . . . that is not linked to current sensory input." While recognizing that, even among human beings, explicit evidence of "internal images" cannot be derived from current techniques of observation, Griffin suggests that at least we can make inferences that can be tested. The *prima facie* case for awareness and consciousness remains strong.

Awareness and consciousness are again subjects of Wilson's *mentalism,* which he calls "the strongest redoubt of counterbiology" (Introduction). Although it may eventually be shown, as Wilson maintains it will, that mental phenomena such as mind and consciousness can be reduced to "conventional neuronal circuitry," the question of how nondeterministic behavior can arise by exclusively deterministic evolutionary processes is a challenge to the theoretical basis of sociobiology.

A third major point of contention is the methodology of sociobiology. The attacks that have been made against the terminological reductionism (for example, the terms *cuckoldry, nepotism,* and *slavery*) are well known and not all groundless. In this volume, Schneewind (Chapter Twelve), among others, criticizes Wilson's undiscriminating use of figurative language in his accounts of animal behaviors. Methodological reductionism is, however, another matter. While Wilson (Introduction), following Karl Popper, argues that such reductionism is necessary for the advancement of science, several of our authors sharply disagree. Hardin (Chapter Nine) says, "Reductionism contributes little or nothing to the understanding of the interactions of organisms." Perhaps the strongest criticism is that of Washburn (Chapter Three), who says of Wilson, "When discussing insects (where he has facts), he does not bring human beings into the discussion. But when discussing human beings (where facts are minimal), insects are used to make critical points." Washburn believes that reductionism is an inescapable consequence of Wilson's direction of inquiry: "By investigating human behavior with the questions and techniques suitable for animals with very simple nervous systems, the whole nature of human behavior is lost." Beach (Chapter Six) widens this criticism to interspecific comparisons in general. "Meaningful comparisons between Species A and Species B simply are not possible until the behavior in question has been analyzed with equal care, objectivity, and precision *in both species,*" he says, adding that "It is fruitless and potentially dangerous to attempt to compare a selected aspect of human and animal behavior when the sole basis for selection is a common label referring to superficial similarities." Given the fact that the sociobiological hypothesis rests on the assumption of evolutionary continuities of behavior from insects to humans, Wilson is thus challenged to produce evidence beyond analogies derived from one species and applied to another.

It has been frequently observed during the debate that sociobiological analogies tend to become sociobiological precursors. For example, the contribution of the sterile female ant to her colony is reminiscent of the contribution made by a maiden aunt to her family. In the latter case, the behavior suggests *altruism,* and that is the term sociobiologists borrow from human behavior and apply (as an analogy) to insect behavior. So far, little harm has been done beyond the solecism of attributing a human characteristic to an ant. But the next step is to read that same *altruism* in evolutionary perspective as a precursor, however distant, of the serviceable spinster's behavior. At this point, great harm may be done by suggesting that the evolutionary derivation of this and other human behaviors has been somehow explained by this extremely curious method of reasoning.

As we have seen, sociobiology has been criticized in this volume on three principal grounds: its inadequate explanation of human culture, its failure to encompass mental phenomena, and the directionality of its evolutionary hypothesis (using simple behaviors to account for complex ones), which necessarily involves reductionism. We ought to remember, however, that the crux of these arguments is the extension of sociobiology into the human realm. For the most part, there has been little quarrel with either hypothesis or data in the application of sociobiological methods of inquiry into nonhuman species, especially at levels below the primates.

Wilson, building on the work of earlier ethologists and population geneticists, has propounded a grand hypothesis unparalleled in biology since 1859. Philosophers and historians of science have interested themselves in the implicit processes involved in the formation and acceptance of hypotheses. In this volume, Hull (Chapter Seven) compares Wilson's hypothesis with Darwin's (against the background of a third hypothesis—phrenology). He concludes, "What really determines the success or failure of new scientific theories is how advocates of these views continue to conduct themselves. They must be conceptually flexible, socially cohesive, and terminologically rigid." Hull finds that the Darwinians successfully met these criteria, while there is substantial doubt about how the sociobiologists will fare in these respects.

There are, of course, vast differences of other kinds between *Sociobiology: The New Synthesis* (1975a) and *On the Origin of Species* ([1859] 1964). For one thing, Darwin was far more cautious in his claims than Wilson has been. There is nothing in Wilson that parallels Darwin's scrupulously methodical consideration of objections and counterhypotheses, and in 1859 Darwin did not extend evolution to the human level. Darwin was almost morbidly sensitive about upsetting the opinions of his family, his friends, and indeed the public. Wilson, by his own account, was astonished at the turmoil his book created in 1975, but he had little reason to be. There was nothing cautious about his announcing in *Sociobiology* the imminent demise of a number of fields (sociology, ethology, comparative psychology, and the humanities) unless they submitted themselves to the "Modern Synthesis" he himself was championing (Wilson, 1975a). Whether or not Darwin's sensitivity for the feelings of others was a strength or weakness in a scientist, Wilson certainly chose not to imitate him. Leaving his opening and closing references to Albert Camus aside for the moment, statements such as the following almost certainly guaranteed Wilson no more than a guarded public reception: "It seems that our autocatalytic social evolution has locked us onto a particular course which the early hominids still within us may not welcome. To maintain the species indefinitely, we are compelled to drive toward total knowledge, right down to the levels of the neuron and gene. When we have progressed enough to explain ourselves in these mechanistic terms and the social sciences come to full flower, the result might be hard to accept" (1975a, p. 575).

Among the authors in this volume who have viewed sociobiology from the perspective of the history of ideas (Schneewind, Hull, and Holton), it is Holton (Chapter Four) who provides a comprehensive summary of the present status of "the new synthesis." Sociobiology is undeniably a daring hypothesis in an age when scientists in nearly all disciplines, except subatomic physics, have become what Thomas Kuhn in his *Structure of Scientific Revolutions* (1962) called "puzzle solvers." It will galvanize the fields challenged to review their own assumptions and methods, and that in itself will have a beneficial effect. And finally, according to Holton, sociobiology itself is an artifact of present value orientations, ethics, and modes of thought. E. O. Wilson and his supporters are notable, says Holton, "in their plea for a sophisticated form of flexible, almost stochastic, predeterminism and materialism; in their apparently dispassionate concern with a secularized ethic; in their accent on rationality and their underemphasis on affect and symbolic forms."

In sum, sociobiology tells us much about the inherited behavior of animals, and it tells us in a new way that we ourselves are animals. But it has little to tell us about what we most need to know, namely our essential humanness. It seems obvious that our ancestral modes of coping with competitors, predators, and territorial invaders are deeply woven into our genetic structure. That is where sociobiology can inform us and help us become aware of gene-driven compulsions toward behaviors that now are clearly suicidal. But in the long run we need to have greater understanding of our uniqueness, and that is where sociobiology signally fails. As Schneewind (Chapter Twelve) observes, the conclusion of Wilson's book is strangely divided in its emphasis. We are told that without "total knowledge" of gene and neuron and the determinism that governs their actions we cannot "maintain the species indefinitely." Yet Wilson also quotes Camus to the effect that such cold rationalism will bring forth a world we cannot inhabit: "A world that can be explained even with bad reasons is a familiar world. But, on the other hand, in a universe divested of illusions and lights, man feels an alien, a stranger. His exile is without remedy since he is deprived of the memory of a lost home or the hope of a promised land" (Camus, 1955, p. 5). Following this passage from Camus is the last line in Wilson's book. I quote it here in deserved italics: *"This, unfortunately, is true. But we still have another hundred years"* (Wilson, 1975a, p. 575).

Wilson chooses to end his book with an enigma. We have a century to unravel the strands of our inheritance as beasts, but in the process we shall have lost all grounds for belief in our humanity. This is indeed an enigma, but it is also a challenge from Wilson, first to understand him and then to try to prove him wrong. We have no recourse but to accept his challenge. And, paradoxically, he deserves our thanks for having cast it in so extreme a form.

References

Alcock, J. *Animal Behavior*. Sunderland, Mass.: Sinauer Associates, 1975.

Alexander, R. D. "The Search for an Evolutionary Philosophy of Man." *Proceedings of the Royal Society of Victoria*, 1971, *84*, 99–120.

Alexander, R. D. "The Evolution of Social Behavior." *Annual Review of Ecology and Systematics*, 1974, *5*, 325–383.

Alexander, R. D. "The Search for a General Theory of Behavior." *Behavioral Science*, 1975, *20*, 77–100.

Alexander, R. D. "Evolution, Human Behavior, and Determinism." In F. Suppe and P. D. Asquith (Eds.), *PSA 1976*. Vol. 2. East Lansing, Mich.: Philosophy of Science Association, 1977.

Alexander, R. D., and Sherman, P. W. "Local Mate Competition and Parental Investment in Insects." *Science*, 1977, *196*, 494–500.

Allen, L., and others. "Sociobiology—Another Biological Determinism." *BioScience*, 1976, *26*, 182–186.

Alper, J., Beckwith, J., and Miller, L. "Sociobiology Is a Political Issue." In A. L. Caplan (Ed.), *The Sociobiology Debate*. New York: Harper & Row, 1978.

Ardrey, R. *The Territorial Imperative*. New York: Atheneum, 1966.

295

Ayala, F. *Molecular Evolution*. San Francisco: W. H. Freeman, 1975.

Barash, D. P. "The Social Behavior of the Olympic Marmot." *Animal Behaviour,* 1973a, *6,* 173–245.

Barash, D. P. "Social Variety in the Yellow-bellied Marmot." *Animal Behaviour,* 1973b, *21,* 579–584.

Barash, D. P. "The Evolution of Marmot Societies: A General Theory." *Science,* 1974a, *185,* 415–420.

Barash, D. P. "Social Behaviour of the Hoary Marmot (*Marmota caligata*)." *Animal Behaviour,* 1974b, *22,* 257–262.

Barash, D. P. "Ecology of Paternal Behavior in the Hoary Marmot (*Marmota caligata*): An Evolutionary Interpretation." *Journal of Mammalogy,* 1975a, *56,* 612–615.

Barash, D. P. "Evolutionary Aspects of Parental Behavior: The Distraction Behavior of the Alpine Accentor, *Prunella collaris.*" *Wilson Bulletin,* 1975b, *87,* 367–373.

Barash, D. P. "Male Response to Apparent Female Adultery in the Mountain Bluebird (*Sialia currucoides*): An Evolutionary Interpretation." *The American Naturalist,* 1976a, *110,* 1097–1101.

Barash, D. P. "Some Evolutionary Aspects of Parental Behavior in Animals and Man." *American Journal of Psychology,* 1976b, *80,* 195–217.

Barash, D. P. *Sociobiology and Behavior*. New York: Elsevier, 1977.

Barash, D. P., Holmes, W. G., and Greene, P. J. "Exact Versus Probabilistic Coefficients of Relationship: Some Implications for Sociobiology." *The American Naturalist,* in press.

Beach, F. A. "Bisexual Mating Behavior in the Male Rat: Effects of Castration and Hormone Administration." *Physiological Zoology,* 1945, *18* (4), 390–402.

Beach, F. A. "Factors Involved in the Control of Mounting Behavior by Female Mammals." In M. Diamond (Ed.), *Perspectives in Reproduction and Sexual Behavior: A Memorial to William C. Young.* Bloomington: Indiana University Press, 1968, pp. 83–131.

Beck, B. B. "Primate Tool Behavior." In R. Tuttle (Ed.), *Socioecology and Psychology of Primates.* The Hague: Mouton, 1975.

Beckwith, J., and Miller, L. "Behind the Mask of Objective Science." *The Sciences,* 1976, *16* (6), 16–19, 29–31.

Bennett, J. *Rationality: An Essay Towards an Analysis.* London: Routledge & Kegan Paul, 1964.

Bennett, J. *Linguistic Behavior.* New York: Cambridge University Press, 1976.

Bentley, D., and Hoy, R. R. "The Neurobiology of Cricket Song." *Scientific American,* 1974, *231,* 34–44.

Berkeley, G. *De Motu.* In T. E. Jessop (Ed.), *A Bibliography of George Berkeley.* New York: Franklin, 1968. (Originally published 1721.)

Bernds, W. P., and Barash, D. P. "Early Termination of Parental Investment: A General Theory for Mammals, Including Humans." In press.

Bernheim, W. A., and Kluger, M. J. "Fever: Effect of Drug-Induced Antipyresis on Survival." *Science,* 1976, *193,* 237–239.

Bertram, B. C. R. "Kin Selection in Lions and Evolution." In P. P. G. Bareson and R. A. Hinde (Eds.), *Growing Points in Ethology.* London: Cambridge University Press, 1976, pp. 281–302.

Bigelow, R. *The Dawn Warriors.* Boston: Little, Brown, 1969.

Birket-Smith, K. *The Eskimos.* (2nd ed.) London: Methuen, 1959.

Bischof, N. "The Biological Foundations of the Incest Taboo." *Social Sciences Information.* The Hague, Netherlands: Mouton Publishers, 1972, *11* (6) pp. 7–36.

Block, N. J., and Dworkin, G. *The IQ Controversy.* New York: Pantheon, 1976.

Blurton-Jones, N. "Ethology, Anthropology, and Childhood." In R. Fox (Ed.), *Biosocial Anthropology.* London: Malaby Press, 1975, pp. 69–72.

Boorman, S. A., and Levitt, P. R. "Group Selection on the Boundary of a Stable Population." *Proceedings of the National Academy of Sciences,* 1972, *69,* 2711–2713.

Boorman, S. A., and Levitt, P. R. "Group Selection on the Boundary of a Stable Population." *Theoretical Population Biology,* 1973, *4,* 85–128.

Borgstrom, G. *The Food and People Dilemma.* Scituate, Mass.: Duxbury Press, 1974.

Boulding, K. "Guilt by Association." *Technology Review,* 1977, *79* (6), 3.

Braverman, H. *Labor and Monopoly Capital.* New York: Monthly Review Press, 1974.

Brown, J. L. "The Evolution of Diversity in Avian Territorial Systems." *Wilson Bulletin,* 1964, *76,* 160–169.

Brown, J. L. "Alternate Routes to Sociality in Jays—with a Theory for the Evolution of Altruism and Communal Breeding." *American Zoologist,* 1974, *14,* 63–80.

Brown, J. L. *The Evolution of Behavior.* New York: Norton, 1975.

Brown v. *Board of Education.* 347 U.S. 483 (1954).

Bullock, K., and Baden, J. "Communes and the Logic of the Commons." In G. Hardin and J. Baden (Eds.), *Managing the Commons.* San Francisco: W. H. Freeman, 1977.

Bunge, M. "Emergence and the Mind." *Neuroscience,* 1977, *2,* 501–509.

Burian, R. "More than a Marriage of Convenience: On the Inextricability of History and Philosophy of Science." *Philosophy of Science,* 1977, *44,* 1–42.

Burkhardt, F. "England and Scotland: The Learned Societies." In T. Glick (Ed.), *The Comparative Reception of Darwinism.* Austin: University of Texas Press, 1974.

Campbell, B. (Ed.). *Sexual Selection and the Descent of Man, 1871–1971.* Chicago: Aldine, 1972.

Campbell, D. T. "On the Conflicts Between Biological and Social Evolution and Between Psychology and Moral Tradition." *American Psychologist,* 1975, *30* (12), 1103–1126.

Camus, A. *The Myth of Sisyphus and Other Essays.* (J. O'Brien, Trans.) New York: Random House, 1955.

Cannon, W. B. *Bodily Changes in Pain, Hunger, Fear, and Rage.* New York: Appleton-Century-Crofts, 1927.

Cantor, G. N. "The Edinburgh Phrenology Debate: 1803–1828." *Annals of Science,* 1975, *32,* 195–218.

Caplan, A. "Ethics, Evolution, and the Milk of Human Kindness." *Hastings Center Report,* 1976, *6,* 20–25.

Cavalli-Sforza, L. L. "The Genetics of Human Populations." *Scientific American,* 1974, *9,* 81–89.

Chagnon, N. A. "Fission in an Amazonian Tribe." *The Sciences* (New York Academy of Science), 1976, *16,* 14–18.

Chagnon, N. A., and Irons, W. (Eds.). *Evolutionary Biology and Human Social Organization.* Scituate, Mass.: Duxbury Press, in press.

Chasin, B. "Sociobiology: A Sexist Synthesis." *Science for the People,* 1977, *9* (3), 27–31.

Chevalier-Skolnikoff, S. "Male-Female, Female-Female, and Male-Male Sexual Behavior in the Stumptail Monkey, with Special

Attention to the Female Orgasm." *Archives of Sexual Behavior,* 1974, *3,* 95–116.

Chomsky, N. *Reflections on Language.* New York: Random House, 1975.

Cloud, W. "After the Green Revolution," *The Sciences,* New York Academy of Sciences, 1973, *13,* 9–11.

Cody, M. L. "Finch Flocks in the Mohave Desert." *Theoretical Population Biology,* 1971, *2,* 142–158.

Cody, M. L. "Optimization in Ecology." *Science,* 1974, *183,* 1156–1164.

Cohen, J. E. *Casual Groups of Monkeys and Men: Stochastic Models of Elemental Social Systems.* Cambridge, Mass.: Harvard University Press, 1971.

Collins, R. L. "On the Inheritance of Handedness, I. Laterality in Inbred Mice." *Journal of Heredity,* 1968, *59,* 9–12.

Cooter, R. J. "Phrenology: The Provocation of Progress." *History of Science,* 1976, *14,* 211–234.

Corner, G. W. *Hormones in Human Reproduction.* Princeton: Princeton University Press, 1942.

Cowan, R. S. "Nature and Nurture: The Interplay of Biology and Politics." *Studies in History of Biology,* 1977, *1,* 133–208.

Crane, J. "Display, Breeding and Relationship of the Fiddler Crabs." *Zoologics,* 1943, *28,* 217–223.

Crook, J. M. "The Evolution of Social Organization and Visual Communication in the Weaver Birds (*Ploceinae*)." *Behaviour,* Supplement 10, 1964.

Crook, J. M. "Social Organization and the Environment." *Animal Behaviour,* 1970, *18,* 197–209.

Crow, J. F. "Mechanisms and Trends in Human Evolution." *Daedalus,* 1961, *90,* 416–431.

Cullen, E. "Adaptations in the Kittiwake to Cliff-Nesting." *Ibis,* 1957, *99,* 275–302.

Curtin, R. A. "Langur Social Behavior and Infant Mortality." *Kroeber Anthropological Society Papers,* 1977, *50,* 27–36.

Curtin, R., and Dolhinow, P. "The Gray Langur Monkey of India: Primate Social Behavior in a Changing World." *American Scientist,* in press.

Darwin, C. R. *On the Origin of Species.* Facsimile of the 1859 edition.

Cambridge, Mass.: Harvard University Press, 1964. (Originally published 1859.)

Darwin, C. R. *On the Origin of Species*. (6th ed.) New York: Macmillan, 1927. (Originally published 1859.)

Darwin, C. R. *The Descent of Man and Selection in Relation to Sex*. (2 vols.) New York: International Publications Service, 1969. (Originally published 1871.)

Darwin, F. *The Life and Letters of Charles Darwin*. New York: Appleton-Century-Crofts, 1899.

Daubeny, C. G. *The London Times*, July 21, 1864, p. 9.

Davis, B. D. "Sociobiology: The Debate Continues." *Hastings Center Report*, 1976, *6* (5), 19.

Dawkins, R. *The Selfish Gene*. New York: Oxford University Press, 1976.

De Chardin, P. T. *The Phenomenon of Man*. New York: Harper & Row, 1961.

Dement, W. C. *Some Must Watch While Some Must Sleep*. San Francisco: W. H. Freeman, 1974.

De Vore, I. (Ed.). *Sociobiology and the Social Sciences*. Chicago: Aldine, in press.

De Vore, I., Trivers, R. L., and Goethak, G. W. *Exploring Human Nature*. Cambridge, Mass.: Educational Development Corporation, 1973.

Dickeman, M. "Female Infanticide and the Reproductive Strategies of Stratified Human Societies: A Preliminary Model." In N. A. Chagnon and W. G. Irons (Eds.), *Evolutionary Biology and Human Social Organization*. Scituate, Mass.: Duxbury Press, in press.

Dobzhansky, T. *Mankind Evolving: The Evolution of the Human Species*. New Haven, Conn.: Yale University Press, 1962.

Dobzhansky, T. "The Myths of Genetic Predestination and of *Tabula Rasa*." *Perspectives in Biology and Medicine*, 1976, pp. 156–170.

Dostoevsky, F. *Notes from Underground* and *The Double*. New York: Penguin, 1972. (Originally published 1864.)

Dupree, A. H. "'Sociobiology' and the Natural Selection of Scientific Disciplines." *Minerva*, 1977, *115* (1), 94–101.

Durham, W. H. "Resource Competition and Human Aggression." Part I: "A Review of Primitive War." *Quarterly Review of Biology*,

1976, *51*, 385–415.

Education Development Center. *Exploring Human Nature*. Cambridge, Mass.: Education Development Center, 1973.

Ehrman, L., and Parsons, P. A. *The Genetics of Behavior*. Sunderland, Mass.: Sinauer Associates, 1976.

Eibl-Eibesfeldt, I. *Ethology: The Biology of Behavior*. New York: Holt, Rinehart and Winston, 1975.

Eisenberg, J. F., Muckenhirn, N. A., and Rudran, R. "The Relation Between Ecology and Social Structure in Primates." *Science*, 1972, *176*, 863–874.

Eldredge, N., and Gould, S. J. "Punctuated Equilibria: An Alternative to Phyletic Gradualism." In T. J. M. Schopf (Ed.), *Models in Paleobiology*. San Francisco: W. H. Freeman, 1972.

Ellegård, A. *Darwin and the General Reader*. Göteborg: Göteborgs Universitets Årsskrift, 1958.

Emlen, J. M. *Ecology: An Evolutionary Approach*. Reading, Mass.: Addison-Wesley, 1972.

Emlen, S. T. "An Alternative Case for Sociobiology." *Science*, 1976, *192*, 736–738.

Emlen, S. T., and Oring, L. W. "Ecology, Sexual Selection and the Evolution of Mating Systems." *Science*, 1977, *197*, 215–223.

Eshel, I. "On the Neighbor Effect and the Evolution of Altruistic Traits." *Theoretical Population Biology*, 1972, *3*, 258–277.

Evans, N. T. *The Comparative Ethology and Evolution of the Sand Wasps*. Cambridge, Mass.: Harvard University Press, 1966.

Feyerabend, P. *Against Method*. London: New Left Books, 1975.

Feynman, R. P., and others. *Feynman Lectures on Physics* Vol. 1: *Mainly Mechanics, Radiation, and Heat*. Reading, Mass.: Addison-Wesley, 1963.

Feynman, R. P., and others. *Feynman Lectures on Physics*. Vol. 2: *Mainly Electromagnetism and Matter*. Reading, Mass.: Addison-Wesley, 1964.

Feynman, R. P., and others. *Feynman Lectures on Physics*. Vol. 3: *Quantum Mechanics*. Reading, Mass.: Addison-Wesley, 1965.

Fishbein, H. D. *Evolution, Development, and Children's Learning*. Pacific Palisades, Calif.: Goodyear, 1976.

Fisher, R. A. *The Genetical Theory of Natural Selection*. Oxford, England: Clarendon, 1930.

Fox, R. (Ed.). *Biosocial Anthropology*. London: Malaby Press, 1975.

Freedman, D. G. *Human Infancy: An Evolutionary Perspective*. Hillsdale, N.J.: L. Erlbaum, 1974.

Freedman, D. G. *Human Sociobiology*. New York: Macmillan, in press.

Freud, S. *Civilization and Its Discontents*. (J. Strachey, Ed. and Trans.) New York: Norton, 1962.

Frisch, K. von. *The Dance Language and Orientation of Bees*. (L. Chadwick, Trans.) Cambridge, Mass.: Harvard University Press, 1967.

Frisch, K. von. *Bees, Their Vision, Chemical Senses, and Language*. (Revised ed.) Ithaca, N.Y.: Cornell University Press, 1971.

Fuller, J. L., and Thompson, W. R. *Behavior Genetics*. (2nd ed.) St. Louis: Mosby, 1978.

Gadgil, M. "Evolution of Social Behavior Through Interpopulation Selection." *Proceedings of the National Academy of Sciences*, 1975, *72*, 1199–1201.

Galton, F. *Hereditary Genius: An Enquiry into Its Laws and Consequences*. New York: Appleton-Century-Crofts, 1884.

Galvin, R. M. "Why You Do What You Do—Sociobiology: A New Theory of Behavior." *Time*, Aug. 1, 1977, pp. 54–63.

Geist, V. "On the Relationship of Social Evolution and Ecology in Ungulates." *American Zoologist*, 1974, *14*, 205–220.

Geschwind, N. (Ed.). "Selected Papers on Language and the Brain." *Boston Studies of Science*. Vol. 16 (Synthese Library: No. 68). Dordrecht, Netherlands: D. Reidel, 1974.

Ghiselin M. "A Radical Solution to the Species Problem." *Systematic Zoology*, 1974, *23*, 536–544.

Glick, T. F. (Ed.). *The Comparative Reception of Darwinism*. Austin: University of Texas Press, 1974.

Globus, G. G., Maxwell, G., and Savodnik, I. *Consciousness and the Brain*. New York: Plenum Press, 1976.

Goldberg, S. *The Inevitability of Patriarchy*. New York: Morrow, 1973.

Goldsby, R. A. *Race and Races*. (2nd ed.) New York: Macmillan, 1977.

Gottlieb, G. "Aspects of Neurogenesis." In G. Gottlieb (Ed.), *Studies on the Development of Behavior and the Nervous System*, Vol. 2. New York: Academic Press, 1973a.

Gottlieb, G. "Behavioral Embryology." In G. Gottlieb (Ed.), *Studies on the Development of Behavior and the Nervous System,* Vol. 1. New York: Academic Press, 1973b.

Gottlieb, G. "Development of Neural and Behavioral Specificity." In G. Gottlieb (Ed.), *Studies in the Development of Behavior and the Nervous System,* Vol. 3. New York: Academic Press, 1976.

Gould, J. L. "Honey Bee Recruitment: The Dance-Language Controversy." *Science,* 1975, *189,* 685–693.

Gould, J. L. "The Dance-Language Controversy." *Quarterly Review of Biology,* 1976, *51,* 212–244.

Gould, S. J. "Biological Potential vs. Biological Determinism." *Natural History,* 1976a, *85* (5), 12–22.

Gould, S. J. "Darwin's Untimely Burial." *Natural History,* 1976b, *85,* 24–30.

Gould, S. J. *Ontogeny and Phylogeny.* Cambridge, Mass.: Harvard University Press, 1977.

Gouldner, A. W. *The Dialectic of Ideology and Technology.* New York: Seabury Press, 1976.

Gregory, F. "Scientific Versus Dialectical Materialism: A Clash of Ideologies in Nineteenth-Century German Radicalism." *Isis,* 1977, *68* (242), 206–223.

Greene, P. "Promiscuity, Paternity and Culture." *American Ethnologist,* in press.

Grene, M. "The Understanding of Nature—Essays in the Philosophy of Biology." *Boston Studies in the Philosophy of Science.* Vol. 23 (Synthese Library No. 66). Dordrecht, Netherlands: D. Reidel, 1974, Chaps. 16–19.

Grice, H. P. "Meaning." *Philosophical Review,* 1957, *66,* 377–388.

Griffin, D. R. *The Question of Animal Awareness.* New York: Rockefeller University Press, 1976.

Habermas, J. *Toward a Rational Society.* Boston: Beacon Press, 1971.

Haeckel, E. *The Riddle of the Universe.* London: Watts, 1929. (Originally published 1899.)

Hailman, J. P. "Breeding Synchrony in the Equatorial Swallow-tailed Gull." *American Naturalist,* 1964, *98,* 79–83.

Haller, J. S. *Outcasts from Evolution.* Urbana: University of Illinois Press, 1971.

Hamburg, D. A. "Emotions in the Perspective of Human Evolution." In P. H. Knapp (Ed.), *Expression of the Emotions in Man.* New York: International Universities Press, 1963, pp. 300–317.

Hamilton, W. D. "The Genetical Evolution of Social Behaviour." *Journal of Theoretical Biology,* 1964, 7, 1–52.

Hamilton, W. D. "Extraordinary Sex Ratios." *Science,* 1967, *156,* 477–488.

Hamilton, W. D. "Geometry for the Selfish Herd." *Journal of Theoretical Biology,* 1971, *31,* 295–311.

Hamilton,, W. D. "Altruism and Related Phenomena, Mainly in Social Insects." *Annual Review of Ecology and Systematics,* 1972, *3,* 193–232.

Hamilton, W. D. "Innate Social Aptitudes of Man: An Approach from Evolutionary Genetics." In R. Fox (Ed.), *Biosocial Anthropology.* London: Malaby Press, 1975, pp. 37–67.

Hardin, G. "The Tragedy of the Commons." *Science,* 1968, *162,* 1243–1248.

Hardin, G. *Stalking the Wild Taboo.* Los Altos, Calif.: Kaufman, 1973.

Hardin, G. *The Limits of Altruism.* Bloomington: Indiana University Press, 1977.

Harris, D. "Enlightenment." *The New Encyclopaedia Britannica,* 1974, *6,* 892.

Harris, M. *Culture, Man, and Nature.* New York: Crowell, 1968.

Hartshorne, C. *Born to Sing, an Interpretation and World Survey of Bird Song.* Bloomington: Indiana University Press, 1973.

Hartung, J. "On Natural Selection and the Inheritance of Wealth." *Current Anthropology,* 1976, *17* (4), 607–622.

Hattiangadi, J. N. "Mind and the Origin of Language." *Philosophical Forum,* 1973, *14,* 81–98.

Heidegger, M. *Sein und Zeit.* Frankfrut: Klostermann, 1927.

Heston, L. L., and Shields, J. "Homosexuality in Twins: A Family Study and a Registry Study." *Archives of General Psychiatry,* 1968, *18,* 149–160.

Himmelfarb, G. *Darwin and the Darwinian Revolution.* New York: Doubleday, 1959.

Hinde, R. A. *Animal Behavior: A Synthesis of Ethology and Comparative Psychology.* New York: McGraw-Hill, 1970.

Hockett, C. P. "Biophysics, Linguistics, and the Unity of Science." *American Scientist,* 1948, *36,* 558–572.

Hodos, W. "Evolutionary Interpretation of Neural and Behavioral Studies of Living Vertebrates." In F. O. Schmitt (Ed.), *The Neurosciences: Second Study Program.* New York: Rockefeller University Press, 1970, pp. 27–38.

Hodos, W., and Campbell, C. B. G. "*Scala naturae:* Why There Is No Theory in Comparative Psychology." In C. N. Cofur (Ed.), *Psychological Review,* 1969, *76* (4), 337–350.

Hoffer, E. *The True Believer.* New York: Harper & Row, 1951.

Hoffman, M. "Homosexuality." In F. A. Beach (Ed.), *Human Sexuality in Four Perspectives.* Baltimore, Md.: Johns Hopkins University Press, 1977.

Hölldobler, B. "Communication by Tandem Running in the Ant *Camponotus sericeus.*" *Journal of Comparative Physiology,* 1974, *90,* 105–127.

Holton, G. *Scientific Imagination: Case Studies.* New York: Cambridge University Press, 1978.

Hoogland, J. L., and Sherman, P. W. "Advantages and Disadvantages of Book Swallow (*Riparia riparia*) Coloniality." *Ecological Monographs,* 1976, *46,* 33–58.

Hooker, R. "Of the Laws of Ecclesiastical Polity." In J. Keble (Ed.), *Works.* Bk. 1. Oxford, England: Oxford University Press, 1845. (Originally published 1594.)

Hopkins, W. "Physical Theories of the Phenomena of Life." *Fraser's,* 1860, *61,* 739–752; *62,* 74–90.

Hrdy, S. B. "Male-Male Competition and Infanticide Among the Langurs (*Presbytis entellus*)." *Folia Primatologie,* 1974, *22,* 19–58.

Hrdy, S. B. "Infanticide as a Primate Reproductive Strategy." *American Scientist,* 1977, *65,* 40–49.

Hull, D. L. *Darwin and His Critics.* Cambridge, Mass.: Harvard University Press, 1973.

Hull, D. L. "Central Subjects and Historical Narratives." *History and Theory,* 1975, *14,* 253–274.

Hull, D. L. "Are Species Really Individuals?" *Systematic Zoology,* 1976, *25,* 174–191.

Hull, D. L. "Sociobiology: Another New Synthesis." In G. W. Barlow and J. Silverberg (Eds.), *Sociobiology: Beyond Nature/Nurture.* Boulder, Colo.: Westview Press, 1978.

Hull, D. L. "A Matter of Individuality." *Philosophy of Science,* in press.

Hume, D. *A Treatise of Human Nature.* L. A. Selby-Bigger, Ed. Oxford: Oxford University Press, 1975. (Originally published in 1739.)

Humphrey, N. K. "Review of *The Question of Animal Awareness.*" *Animal Behaviour,* 1977, *25,* 521–522.

Hutt, C. *Males and Females.* Middlesex, England: Penguin Books, 1972.

Huxley, J. S. *Evolution: The Modern Synthesis.* London: Allen & Unwin, 1942.

Huxley, L. *Life and Letters of Sir Joseph Dalton Hooker.* London: Murray, 1918.

Irons, W. "Emic and Reproductive Success." In N. A. Chagnon and W. Irons (Eds.), *Evolutionary Biology and Human Social Organization.* Scituate, Mass.: Duxbury Press, in press.

Jacobsen, M. *Developmental Neurobiology.* New York: Holt, Rinehart and Winston, 1970.

Jarman, P. J. "The Social Organization of Antelope in Relation to Their Ecology." *Behaviour,* 1974, *58,* 215–267.

Jaynes, J. *The Origin of Consciousness in the Breakdown of the Bicameral Mind.* Boston: Houghton Mifflin, 1977.

Jensen, A. R. "How Much Can We Boost IQ and Scholastic Achievement?" *Harvard Educational Review,* 1969, *39,* 1–123.

Johannsen, W. *Elemente der exakten Erblichkeitslehre.* Jena, Germany: Gustav Fisher, 1909.

Joravsky, D. *The Lysenko Affair.* Cambridge, Mass.: Harvard University Press, 1970.

Joravsky, D. "Calls for Transformation." *Science,* 1977, *197,* 246–247.

Kamin, L. J. *The Science and Politics of IQ.* Potomac, Md.: Lawrence Erlbaum Associates, 1974.

Katchadourian, H. A., and Lunde, D. T. *Fundamentals of Human Sexuality.* New York: Holt, Rinehart and Winston, 1972.

Katz, P. L. "A Long-Term Approach to Foraging Optimization." *American Naturalist,* 1974, *108,* 758–782.

Keeton, W. T. *Biological Science.* (2nd ed.) New York: Norton, 1972.

Kenny, A. J. P., and others. *The Nature of Mind.* Edinburgh: Edinburgh University Press, 1972.

Kenny, A. J. P., and others. *The Development of Mind.* Edinburgh: Edinburgh University Press, 1973.

Kessel, E. L. "The Mating Activities of Balloon Flies." *Systematic Zoology,* 1955, *4,* 97–104.

King, M. C., and Wilson, A. C. "Evolution at Two Levels in Humans and Chimpanzees." *Science,* 1975, *188,* 107–116.

Kinsey, A. C., Pomeroy, W. B., and Martin, C. E. *Sexual Behavior in the Human Male.* Philadelphia: Saunders, 1948.

Konner, M. J. "Aspects of the Developmental Ethology of a Foraging People." In N. Blurton-Jones (Ed.), *Ethological Studies of Child Behavior.* Cambridge, England: Cambridge University Press, 1972.

Krebs, H. A. "Review of *The Question of Animal Awareness.*" *Nature,* 1977, *266,* 792.

Kroodsma, D. E. "Reproductive Development in a Female Songbird: Differential Stimulation by Quality of Male Song." *Science,* 1977, *192,* 574–575.

Kruuk, H. "Predators and Anti-Predator Behaviour of the Blackheaded Gull (*Larus ridibandus*)." *Behaviour,* 1964, Supplement 11.

Kuhn, T. S. *The Structure of Scientific Revolutions.* Chicago: University of Chicago Press, 1962.

Kuhn, T. S. *The Structure of Scientific Revolutions.* (2nd ed., Enlarged.) Chicago: University of Chicago Press, 1970.

Kung, H. *On Being a Christian.* Garden City, N.Y.: Doubleday, 1976.

Larsen, R. R. "On Comparing Man and Ape: An Evaluation of Methods and Problems." *Man,* 1976, *11,* 202–219.

Lashley, K. S. "The Behavioristic Interpretation of Consciousness." *Psychological Reviews,* 1923, *30,* 237–272, 329–353.

Layzer, D. "Heritability Analyses of IQ: Science or Numerology?" *Science,* 1974, *183,* 1259–1266.

Lehrman, D. E. "Semantics and Conceptual Issues in the Nature-Nurture Problem." In L. R. Aronson and others (Eds.), *Development and Evolution of Behavior.* San Francisco: W. H. Freeman, 1970.

Lenski, G., and Lenski, J. *Human Societies: An Introduction to Macrosociology.* New York: McGraw-Hill, 1978.

Levins, R. "The Strategy of Model Building in Population Biology." *American Scientist,* 1966, *54,* 421–431.

Levins, R. *Evolution in Changing Environments.* Princeton, N.J.: Princeton University Press, 1968.

Levins, R. "Extinction." In M. Gerstenhaber (Ed.), *Some Mathematical Questions in Biology,* Vol. 1. Providence, R.I.: American Mathematical Society, 1970.

Levins, R. "Evolution in Communities Near Equilibrium." In M. L. Cody and J. R. Diamond (Eds.), *Ecology and Evolution of Communities.* Cambridge, Mass.: Harvard University Press, 1975, pp. 16–50.

Lewin, R. "The Course of a Controversy." *The New Scientist,* May 13, 1976, *70,* 344–345.

Lewontin, R. C. "The Apportionment of Human Diversity." *Evolutionary Biology,* 1972, *6,* 381–398.

Limbaugh, C. "Cleaning Symbiosis." *Scientific American,* 1961, *205* (2), 42–50.

Lin, N., and Michener, C. D. "Evolution of Sociality in Insects." *Quarterly Review of Biology,* 1972, *47,* 131–159.

Lindauer, M. "Schwarmbienen auf Wohnungssuch." *Zeitschrift für Vergleichende Physiologie,* 1955, *37,* 263–324.

Lindauer, M. *Communication Among Social Bees.* (rev. ed.) Cambridge, Mass.: Harvard University Press, 1971.

Lister, M. *A Journey to Paris in the Year Sixteen Ninety-Eight.* (R. P. Stearns, Ed.) Urbana: University of Illinois Press, 1967. (Originally published 1699.)

Loeb, J. *The Mechanistic Conception of Life.* Cambridge, Mass.: Harvard University Press, 1964. (Originally published 1912.)

Loehlin, J. C., and Nichols, R. C. *Heredity, Environment, and Personality.* Austin: University of Texas Press, 1976.

London Times. April 8, 1871, p. 5.

Lorenz, K. *King Solomon's Ring.* New York: Crowell, 1952.

Lorenz, K. "The Evolution of Behavior." *Scientific American,* 1958, *199,* 67–78.

Lorenz, K. *On Aggression.* New York: Harcourt Brace Jovanovich, 1963.

Lorenz, K. *On Aggression.* New York: Harcourt Brace Jovanovich, 1966.

Lucretius. *The Nature of the Universe.* Book I, line 25. Chicago: Regnery, 1969. [c. 99–55 B.C.]

Macalister, A. "Phrenology." *Encyclopaedia Britannica,* 1888, *18,* 842–849.

Mach, E. *The Science of Mechanics.* (T. J. McCormack, Trans.) LaSalle, Ill.: Open Court, 1974.

Mackenzie, B. D. *Behaviourism and the Limits of Scientific Method.* London: Routledge & Kegan Paul, 1977.

McKusick, V. A., and Ruddle, F. H. "The Status of the Gene Map of the Human Chromosome." *Science,* 1977, *196,* 390–405.

Malinowski, B. *A Scientific Theory of Culture and Other Essays.* Chapel Hill: University of North Carolina Press, 1944.

Malthus, T. R. *Population: The First Essay.* Ann Arbor: University of Michigan Press, 1959. (Originally published 1798.)

Marshall, A. J. *Bower-birds, Their Displays and Breeding Cycles.* London: Oxford University Press, 1954.

Marx, K. "Critique of the Gotha Program." In R. C. Tucker (Ed.), *Marx-Engels Reader.* New York: Norton, 1972.

Masters, R. D. "The Implications of Sociobiology." *Science,* 1976, *192,* 427–428.

Maynard Smith, J. "Group Selection and Kin Selection." *Nature,* 1964, *201* (4924), 1145–1147.

Maynard Smith, J. "The Evolution of Alarm Calls." *American Naturalist,* 1965, *99,* 59–63.

Maynard Smith, J. "What Use Is Sex?" *Journal of Theoretical Biology,* 1971, *30,* 319–335.

Maynard Smith, J. "Group Selection." *Quarterly Review of Biology,* 1976, *51,* 277–283.

Maynard Smith, J., and Price, G. R. "The Logic of Animal Conflict." *Nature,* 1973, *246,* 15–18.

Maynard Smith, J., and Ridpath, M. G. "Wife-Sharing in the Tasmanian Native Hen, *Tribonyx mortierii:* A Case of Kin Selection?" *American Naturalist,* 1972, *106,* 447–452.

Mayr, E. "Behavior Programs and Evolutionary Strategies." *American Scientist,* 1974, *62,* 650–659.

Mazur, A. "A Cross-Species Comparison of Status in Small Established Groups." *American Sociological Review*, 1973, *38*, 513–530.

Medvedev, Z. A. *The Rise and Fall of T. D. Lysenko.* New York: Columbia University Press, 1969.

Mendelsohn, E. "Revolution and Reduction." In Y. Elkana (Ed.), *The Interaction Between Science and Philosophy.* New York: Academic Press, 1974.

Merleau-Ponty, M. *La Phenomenologie de la Perception.* Paris: 1945.

Möglich, M., Maschwitz, U., and Hölldobler, B. "Tandem Calling: A New Kind of Signal in Ant Communication." *Science*, 1974, *186*, 1046–1047.

Mohnot, S. M. "Some Aspects of Social Changes and Infant Killing in the Hanuman Langur (*Presbytis entellus*) in Western India." *Mammalia*, 1971, *35*, 195–198.

Morison, R. S. "Sociobiology: The Debate Continues." *Hastings Center Report*, 1976, *6* (5), 18–19.

Myers, R. E. "Comparative Neurology of Vocalization and Speech: Proof of a Dichotomy." *Annals of New York Academy of Sciences*, 1976, *280*, 745–757.

Nagel, T. *The Possibility of Altruism.* Oxford, England: Oxford University Press, 1970.

Naroll, R., and Divale, W. T. "Natural Selection in Cultural Evolution: Warfare Versus Peaceful Diffusion." *American Ethnologist*, 1976, *3*, 97–129.

Niebuhr, R. *Beyond Tragedy.* New York: Scribner's, 1937.

Noble, D. F. *America by Design.* New York: Knopf, 1977.

Ophuls, W. *Ecology and the Politics of Scarcity.* San Francisco: W. H. Freeman, 1977.

Orians, G. N. "The Ecology of Blackbird (*Agelaius*) Social Systems." *Ecological Monographs*, 1961, *31*, 285–312.

Orians, G. N. "On the Evolution of Mating Systems in Birds and Mammals." *American Naturalist*, 1969, *103*, 589–603.

Packer, C. "Reciprocal Altruism in *Papio Anubis*." *Nature*, 1977, *265*, 441–443.

Panati, C., and Monroe, S. "Shockley Revisited," *Newsweek*, April 12, 1976, pp. 51–52.

Parker, S. "The Precultural Basis of the Incest Taboo." *American Anthropologist*, 1976, *73* (2), 285–305.

Parssinen, T. M. "Popular Science and Society: The Phrenology Movement in Early Victorian Britain." *Journal of Social History,* 1974, *8,* 1–20.

Patterson, I. J. "Timing and Spacing of Broods in the Black-headed Gull, *Larus ridibundus." Ibis,* 1965, *107,* 433–459.

Pauling, L. *General Chemistry.* (3rd ed.) San Francisco: W. H. Freeman, 1970.

Peckham, M. "Darwinism and Darwinisticism." *Victorian Studies,* 1959, *3,* 19–40.

Peters, R. H. "Tautology in Evolution and Ecology." *American Naturalist,* 1976, *110,* 1–12.

Pianka, E. R. *Evolutionary Ecology.* New York: Harper & Row, 1974.

Plessner, H. *Die Stufen des Organischen und der Mensch.* Berlin: De Gruyer, 1928.

Plessner, H. *Philosophische Anthropologie.* Frankfurt: Fischer, 1970.

Popper, K. R. *The Open Society and Its Enemies.* (5th ed.) Princeton, N.J.: Princeton University Press, 1966.

Popper, K. R. "Scientific Reduction and the Essential Incompleteness of All Science." In F. J. Ayala and T. Dobzhansky (Eds.), *Studies in the Philosophy of Biology.* Berkeley: University of California Press, 1974.

Pryor, K. *Lads Before the Wind.* New York: Harper & Row, 1976.

Pugh, G. *The Biological Origin of Human Values.* New York: Basic Books, 1977.

Roget, P. M. *Animal and Vegetable Physiology, Considered with Reference to Natural Theology* (Bridgewater Treatise). London: Pickering, 1834.

Roget, P. M. "Phrenology." *Encyclopaedia Britannica,* 1842, *17,* 454–473.

Roget, P. M. *Thesaurus of English Words and Phrases.* New York: St. Martin's, 1965. (Originally published 1852.)

Rose, H., and Rose, S. *The Radicalisation of Science.* London: Macmillan, 1976.

Rüppel, G. "Eine 'Lüge' als Gerichtete Mitteilung beim Eisfuchs (*Alopex lagopus L.*)." *Zeitschrift für Tierpsychologie,* 1969, *26,* 371–374.

Ruse, M. "The Nature of Scientific Models: Formal vs. Material Analogy." *Philosophy of the Social Sciences,* 1973, *3,* 63–80.

Ruse, M. "Charles Darwin and Artificial Selection." *Journal of the History of Ideas,* 1975a, *36,* 339–350.

Ruse, M. "Darwin's Debt to Philosophy." *Studies in the History and Philosophy of Science,* 1975b, *6,* 159–181.

Sahlins, M. *Culture and Practical Reason.* Chicago: University of Chicago Press, 1976a.

Sahlins, M. *The Use and Abuse of Biology.* Ann Arbor: University of Michigan Press, 1976b.

Salzman, F. "Are Sex Roles Biologically Determined?" *Science for the People,* 1977, *9* (4), 27–32.

Schrödinger, E. *What Is Life?* New York: Cambridge University Press, 1967.

Schuster, C. "Knowledge and Beyond Knowledge in Political Philosophy." Address to Northwest Philosophy Conference, Central Washington State University, Ellensburg, Washington, November 18–19, 1977.

Scott, J. P. "Introduction." *Minutes of the Conference on Genetic and Social Behavior.* Bar Harbor, Maine: R. B. Jackson Memorial Laboratory, 1946.

Scott, J. P. Foreword to "Methodology and Techniques for the Study of Animal Societies." *Annals of the New York Academy of Sciences,* 1950, *51,* 1003–1005.

Scrope, G. P. "Combe's 'Outlines of Phrenology' (1836)." *Quarterly Review,* 1836, *57,* 169–182.

Searle, J. R. *Speech Acts.* New York: Cambridge University Press, 1969.

Sebeok, T. A. (Ed.). *How Animals Communicate.* Bloomington: Indiana University Press, 1977.

Sedgwick, A. "Review of *Vestiges of the Natural History of Creation* (1844)." *The Edinburgh Review,* 1845, *82,* 1–45.

Sedgwick, A. "Objections to Mr. Darwin's Theory of the Origin of Species." *The Spectator,* April 7, 1860, pp. 334–335.

Selye, H. *Stress Without Distress.* Philadelphia: Lippincott, 1974.

Shapin, S. "Phrenological Knowledge and the Social Structure of Early 19th-Century Edinburgh." *Annals of Science,* 1975, *32,* 219–243.

Shepher, J. "Mate Selection Among Second-Generation Kibbutz Adolescents and Adults." *Archives of Sexual Behavior,* 1971, *1* (4), 293–307.

Sherman, P. "Nepotism and the Evolution of Alarm Calls." *Science,* 1977, *197,* 1246–1253.

Sherrington, C. *Man on His Nature.* Cambridge, England: Cambridge University Press, 1940.

Simpson, G. G. *The Meaning of Evolution.* New Haven, Conn.: Yale University Press, 1949.

Simpson, G. G. *Principles of Animal Taxonomy.* New York: Columbia University Press, 1961.

Simpson, G. G., and Beck, W. S. *Life: An Introduction to Biology,* (2nd Ed.) New York: Harcourt Brace Jovanovich, 1965.

Skinner, B. F. *The Behavior of Organisms.* New York: Appleton-Century-Crofts, 1938.

Skinner, B. F. *Verbal Behavior.* Englewood Cliffs, N.J.: Prentice-Hall, 1957.

Skinner, B. F. "The Phylogeny and Ontogeny of Behavior." *Science,* 1966, *153,* 1205–1215.

Skinner, B. F. *Beyond Freedom and Dignity.* New York: Knopf, 1971.

Skinner, B. F. *About Behaviorism.* New York: Random House, 1974.

Smith, W. J. *The Behavior of Communicating.* Cambridge, Mass.: Harvard University Press, 1977.

Snyder, F. "Toward an Evolutionary Theory of Dreaming." *American Journal of Psychiatry,* 1966, *123,* 121–136.

"Sociobiology: Doing What Comes Naturally." A film distributed by Document Associates, New York, 1976.

Sociobiology Study Group of Science for the People. "Sociobiology—Another Biological Determinism." *BioScience,* 1976, *26,* 182–186.

Sociobiology Study Group of Science for the People. "Sociobiology: A New Biological Determinism." In Ann Arbor Science for the People (Ed.), *Biology as a Social Weapon,* Minneapolis: Burgess, 1977.

Spencer, H. *Principles of Ethics.* New York: Appleton-Century-Crofts, 1896.

Spencer, H. *Autobiography.* London: Williams and Norgate, 1904.

Spinoza, B. *Ethics.* (W. H. White, Trans.; J. Gutman, Ed.) New York: Hafner, 1949. (Originally published 1677.)

Sugiyama, Y. "Social Organization in Hanuman Langurs." In S. A. Altman (Ed.), *Social Communication Among Primates.* Chicago: University of Chicago Press, 1967, pp. 221–236.

Suppe, F., and Asquith, P. D. (Eds.). *PSA 1976.* Vol. 2: *Symposia.* East Lansing, Mich.: Philosophy of Science Association, 1977.

Thorpe, W. H. *Animal Nature and Human Nature.* New York: Doubleday, 1974.

Tiger, L., and Fox, R. *The Imperial Animal.* New York: Dell, 1972.

Tiger, L., and Shepher, J. *Women in the Kibbutz.* New York: Harcourt Brace Jovanovich, 1975.

Tinbergen, N. "Comparative Studies of the Behavior of Gulls (*Laridae*): A Progress Report." *Behaviour,* 1959, *15*, 1–70.

Tinbergen, N. *The Herring Gull's World.* New York: Basic Books, 1960.

Tinbergen, N. "The Shell Menace." *Natural History,* 1963, *72*, 28–35.

Tinbergen, N. "Adaptive Features of the Black-headed Gull *Larus ridibundus.*" *Proceedings of the Fourteenth International Ornithological Congress.* Oxford, England: 1967.

Trivers, R. L. "The Evolution of Reciprocal Altruism." *Quarterly Review of Biology,* 1971, *46* (4), 35–57.

Trivers, R. L. "Parental Investment and Sexual Selection." In B. Campbell (Ed.), *Sexual Selection and the Descent of Man.* Chicago: Aldine, 1972.

Trivers, R. L. "Parent-Offspring Conflict." *American Zoologist,* 1974, *14* (1), 249–264.

Trivers, R. L., and Hare, H. "Haplodiploidy and the Evolution of the Social Insects." *Science,* 1976, *191*, 249–263.

Trivers, R. L., and Willard, D. E. "Natural Selection of Parental Ability to Vary the Sex Ratio of Offspring." *Science,* 1973, *179*, 90–92.

Tucker, R. C. *The Marx-Engels Reader.* New York: Norton, 1972.

Van den Berghe, P. L. "Bringing Beasts Back In." *American Sociological Review,* 1974, *39* (6), 777–788.

Van den Berghe, P. L., and Barash, D. P. "Inclusive Fitness and Human Family Structure." *American Anthropologist,* 1977, *79*, 809–823.

Wade, N. "Sociobiology: Troubled Birth for a New Discipline." *Science,* 1976, *191*, 1151–1155.

Walras, L. (W. Jaffe, Trans.) *Elements of Pure Economics.* Homewood, Ill.: Allen & Unwin, 1954.

Washburn, S. L. "Evolution is a Magic Word." *American Psychologist,* 1976, *31*, 353–355.

Washburn, S. L., and De Vore, I. "Social Behavior of Baboons and Early Man." *Viking Fund Publication in Anthropology,* 1961, *31,* 91–105.

Washburn, S. L., Hamburg, D. A., and Bishop, N. H. "Social Adaptation in Nonhuman Primates." In G. V. Coelho, D. A. Hamburg and J. E. Adams (Eds.), *Coping and Adaptation.* New York: Basic Books, 1974, pp. 3–12.

Weinrich, J. D. "Human Reproductive Strategy." Unpublished doctoral dissertation, Harvard University, 1977.

Weisskopf, V. "Of Atoms, Mountains, and Stars: A Study in Qualitative Physics." *Science,* 1975, *187,* 605–612.

Wells, K. D. "Sir William Lawrence (1783–1867): A Study of Pre-Darwinian Ideas on Heredity and Variation." *Journal of the History of Biology,* 1971, *4,* 319–361.

West-Eberhard, M. J. "The Evolution of Social Behavior by Kin Selection." *Quarterly Review of Biology,* 1975, *50,* 1–33.

Westfall, R. S. "Newton and the Fudge Factor." *Science,* 1973, *179,* 751–758.

Wiens, J. A. "On Group Selection and Wynne-Edwards' Hypothesis." *American Scientist,* 1966, *54,* 273–287.

Williams, G. C. *Natural Selection and Adaptation.* Princeton, N.J.: Princeton University Press, 1966.

Williams, G. C. *Group Selection.* Chicago: Aldine, 1971.

Williams, G. C. *Sex and Evolution.* Princeton, N.J.: Princeton University Press, 1975.

Wilson, E. O. *The Insect Societies.* Cambridge, Mass.: Belknap Press of Harvard University Press, 1971.

Wilson, E. O. *Sociobiology: The New Synthesis.* Cambridge, Mass.: Harvard University Press, 1975a.

Wilson, E. O. "Human Decency Is Animal." *The New York Times Magazine,* October 12, 1975b, 39–50.

Wilson, E. O. "Academic Vigilantism and the Political Significance of Sociobiology." *BioScience,* 1976a, *26,* 183–190.

Wilson, E. O. *Foreword.* In D. P. Barash, *Sociobiology and Behavior.* New York: Elsevier, 1976b.

Wilson, E. O. "Biology and the Social Sciences." *Daedalus,* 1977, *106* (4), 127–140.

Wright, S. *Evolution and the Genetics of Populations.* Vol. 2: *The Theory of Gene Frequencies.* Chicago: University of Chicago Press, 1969.

Wynne-Edwards, V. C. *Animal Dispersion in Relation to Social Behavior.* New York: Hafner, 1962.

Young, R. M. "The Functions of the Brain, Gall to Ferrier." *Isis,* 1968, *59,* 251–268.

Young, R. M. "Darwin's Metaphor: Does Nature Select?" *Monist,* 1971, *55,* 442–503.

Index